D0810682

A MIRROR OF ENGLAND

H. J. Massingham

A Mirror of England

an anthology of the Writings of
H.J. MASSINGHAM
(1888-1952)

'Broadly speaking my theme is the relationship between man and nature in our own country, its fruitfulness and the disastrous consequences of disturbing it.'
Through the Wilderness

edited by
EDWARD ABELSON

with a foreword by
JOHN MICHELL

GREEN BOOKS

This edition published in 1988 by
GREEN BOOKS
Ford House, Hartland
Bideford, Devon EX39 6EE

Cover design by Thomas Keenes and Simon Willby
Photograph of H.J. Massingham by kind permission of Mrs Massingham

Typeset by Computype, Exeter

Printed by Robert Hartnoll (1985) Ltd
Victoria Square, Bodmin, Cornwall

British Library Cataloguing in Publication Data

Massingham, H. J. (Harold John), *1888-1952*
A mirror of England. – (Green classics).
1. England. Countryside, 1901-
I. Title II. Abelson, Edward
942.082

ISBN 1-870098-17-X

CONTENTS

PART THREE 1940-52

FOREWORD

The Posthumous Life of John Massingham

Since the Industrial Revolution we have been living more and more recklessly upon capital, dissipating both the natural wealth of the earth and the inherited wealth of human culture. Most people, even those who do not care to think about it, realize by now that things have gone badly wrong. Even without the nuclear menace, portents of disaster, of extinction even, are apparent everywhere: in the poisoned state of lands, waters and atmosphere, the loss of animal and plant species, deforestation, the depletion of natural resources, aberrant weather patterns, the encroachment of deserts. . . . The list is long, dreary, and deadly. And parallel to these processes is another, also deadly though less easy to evaluate – the decline of regional traditions and culture. With the disappearance of local lore, skills, customs, entertainments, and initiatives an ancient heritage has been lost perhaps irretrievably. It is not such a definable loss as, for example, the dodo or the great auk, for it is not so much material as spiritual. It is a loss which Oliver Goldsmith summed up in two lines of 'The Deserted Village':

> But a bold peasantry, their country's pride,
> When once destroy'd, can never be supplied.

H. J. Massingham was plagued throughout his career by this word, peasantry. By Massingham's time, Goldsmith's term for a body of independent, landed proprietors had been perverted from its natural meaning and become applied to the lowest of social grades, the clod, hick, bumpkin, or rustic idiot. This was a constant hindrance to Massingham's advocacy of the peasant economy, as Goldsmith described it.

> A time there was, ere England's griefs began,
> When every rood of ground maintain'd its man;
> For him light labour spread its wholesome store,
> Just gave what life requir'd, but gave no more;
> His best companions, innocence and health,
> And his best riches, ignorance of wealth.

Massingham's lifetime coincided with the rise and apparent triumph of the forces which he most relentlessly opposed. He saw the village craftsman driven out by mass production, the small farmer replaced by agricultural corporations, rural communities diminished in population, wealth, and spirit through competition from the great urban centres. He witnessed the defacement of England's greatest work of art, its beautiful, productive landscape, by roads, quarries, airports, factories and housing estates, commercial agriculture and forestry. These developments were presented as progress, their destructive aspects being the necessary price for the advancement of civilization. To Massingham they were evidence of cultural and economic decline. Every acre which passed out of cultivation, every family drawn away from the rural economy into urban employment marked a further stage in England's retreat from self-sufficiency into the perils of dependency upon an uncertain world economy.

Like William Cobbett, who once held a village feast to celebrate the failure of a local bank, Massingham knew that one day the tide would turn. This, he said, was not merely desirable but inevitable. A radical change in official attitudes to land and husbandry had to come about, and Massingham urged that it be soon rather than too late. In numerous books and writings he described the beauty of England under the traditional economy and gave encouragement to all who attempted its restoration.

A movement which gave him hope for the future was that initiated by Rolf Gardiner. In the early 1930s Gardiner acquired an estate at Springhead, near Shaftesbury in Dorset, and returned it to mixed cultivation by traditional methods. Organically fertilized and with its own mills and dairy, the farm prospered and expanded. Cottages and workshops were provided for rural craftsmen, and old festivals were renewed to celebrate the various stages of the agricultural year. Gardiner intended by his example to instigate an English renaissance, rooted in the English countryside. Springhead was to be the site of a new university for the study of traditional crafts and methods of organic, co-operative farming. There was wide support for the project, but no government agency or college would make the necessary grants.

Massingham's symposium of 1941, *England and the Farmer,* which he named as his favourite out of all his many books, contained an article by Gardiner. In it he referred to ideas, very similar to his own, which were being carried out at the time in Germany. As a whole-hearted English patriot, unconcerned with international or party politics, Massingham had no particular interest in the German peasant movement. The totalitarian state in which it became submerged stood for everything he most detested. He rejected every system which exalted the state at the expense of local independence and autarky. Thus, despite his admiration for the country radicals at the base of the English socialist tradition, he had no sympathy with the urban, state-centred character of modern socialism. Nor was he in the political sense a conservative. 'He had long realized that the political Right Wing represented an industry, a commerce and a finance he abhorred and blamed', said Edward Hyams in a book which he wrote in partnership with Massingham in 1952. The book, published the following year, after Massingham's death, was called *Prophecy of Famine.* The authors pointed out that Britain had become dangerously dependent upon food imported from abroad; that native resources, talents, and the potential wealth of the land were neglected or wasted; that foreign commodities would not always be cheap and readily available and that Britain could soon be faced with starvation. The remedy, both authors agreed, was an immediate reversal of government policy, directing it towards national self-sufficiency. This could be achieved without any radical innovations, simply by restoring the traditional English economy based on village communities of cultivators and craftsmen. Hyams, an agrarian socialist in the native rather than the Marxian tradition, advocated the nationalization of land, which (as Penelope Massingham pointed out in her preface to the book) H.J.M. would certainly not have supported.

The scope of Massingham's interests is well illustrated in these pages. At the heart of all he wrote was his deep love of England and the products of English culture, in poetry, painting, vernacular building, and the practical crafts. He had no time for the affectations of the arts-and-crafts, ruralist movements which flourished during his earlier years, nor did he admire the 'nighmare forms' of Picasso or the introverted spirit of modern art. Constable's landscapes, the Dorset poetry of William Barnes, Avebury stone circle, a stone-built, well-thatched, working Cotswold village; such characteristic expressions of the English genius were his constant favourites. His eye for country and the poetic quality in his descriptions of natural scenery were similar to Cobbett's, but he was entirely free of the blustering demagogic tendencies in his great hero. Apart from that, he is justly claimed as Cobbett's natural successor.

In his account of the famous 'Rider of the Shires' (included in this anthology), Massingham points out that Cobbett's career ended in apparent failure. So, it could equally well be said, did Massingham's. As indicated in the first paragraph of this essay, the processes which he fought against are still in the ascendency, have increased in power worldwide and now present a vaster, more immediate threat than ever before. The 'Thing', Cobbett's name for the monopolizing system of government, usury, and commerce, outlived him and it outlived Massingham also. Yet at the same time a corresponding reaction has developed. Goldsmith's solitary plaint against rural dereliction has become a swelling chorus, culminating in the modern ecology movement and the emergence of political 'green' parties. The word ecology appeared late in Massingham's life. He understood it as 'the modern equivalent of the old-fashioned idea of good husbandry', applied to the whole earth. Had he lived to the present day, he would still be promoting the idea of ecology, but it is doubtful that he would have joined a green party. The shifts and compromises of politics, and the adoption by green politicians of other causes such as feminism and pacifism, would have deterred him from such associations.

It remains to be seen whether Massingham, and indeed Cobbett, did actually fail as prophets and reformers. As Edward Abelson remarks later, Massingham today is all but forgotten. Many of his views are still considered heretical or outmoded; his archaeological writings, wherein he proposed an Egyptian or Cretan origin for English civilization, are firmly out of favour; there is no time in most people's lives to enjoy his beautiful descriptions of the old England he saw passing. Yet the death of an important writer is often followed by a period of neglect. In Massingham's case, this anthology marks the end of that period. What he said was direct, simple, principled, and never more timely than now. A popular audience, which he never quite achieved in his lifetime, has been created for him, as he knew it would, by the necessity of the times. The voice of informed authority and true orthodoxy is keenly in demand, and Massingham is uniquely qualified to supply it. He was the most unassuming of prophets, a quiet-mannered English country gentleman who would never consciously hurt another soul yet was utterly resolute in standing by what he knew to be right. Through the first half of the twentieth century wasteland he was the firm link in a chain of tradition which has its roots in prehistoric times and ultimately in human nature itself. That tradition – with or without intervening catastrophes – will ultimately reassert itself, and it will become evident, as Arthur Bryant said, that Massingham's philosophy was far deeper and better founded than the systems of Bertrand Russell and the other intellectual lights of his time. Whether or not Massingham will be acclaimed – and he would have been quite indifferent on that score – his vision will endure as long as the human race itself. If it is to be of long endurance, it will be because the traditional world-view, as expressed by John Massingham and his like, has duly prevailed.

John Michell

TRIBUTE TO H. J. MASSINGHAM

By the death of H. J. Massingham the other day England lost a far greater writer and Englishman than is at present realized. Those who take their stand against the popular movements of their time seldom receive recognition during their lives. The more powerful their genius, the louder and harsher the clamour against them. The reward for Socrates' wisdom was a cup of hemlock. John Massingham in his latter years was meted out what must have seemed to him almost as bitter a cup. The truths for which he pleaded and in which he so passionately believed were derided by contemptuous critics, and he himself was termed a sentimentalist and an escapist. The distress he suffered, not from personal neglect – for he was the least vain of men – but from the disregard by his age of all the deeper values in which he believed – was accentuated because he was crippled and, during the last decade of his life, in almost constant pain and ill-health. For the sake of his beliefs he was ready to sacrifice everything, but it would be an exaggeration to say that he endured gladly. He felt every barb acutely.

Yet, under an occasional irritableness and impatience which arose from continual physical suffering and an extreme sensitiveness, he was the gentlest and tenderest-hearted of creatures. For the last decade of his life, drawn to him partly by common beliefs, and partly by the charm and sincerity of his nature, I enjoyed the privilege and happiness of his friendship, and I never met a man more inherently lovable. It was as though the things he loved – and he loved them with childlike and passionate intensity – had become part of his nature: the flowers, fruits and vines he tended, the English countryside he described with such deep understanding and, at times, lyrical inspiration; the craftsmen and husbandmen, skilful, humble, and honourable whom he so fervently championed; the gentle, shaggy sheepdogs in whose fidelity and sagacity he so delighted. There was in him so strong a perception of what constituted goodness in man, beast, and plant and so deep a despair at its lack, that at times his heart could scarcely contain what he felt, and his gnarled, crab-apple, pain-lined face became suffused with the strength of his admiration or anger. But for the never-failing love and care of his wife, who shared all his journeys and battles, his tortured body could never have survived so long as it did the strain of continuous work and the intensity of his feelings; his survival during these last years was a continuous triumph of spirit over matter. So was the campaign he conducted on behalf of what seemed to the overwhelming majority of his contemporaries, and at times to himself, a lost cause.

But it was not a lost cause. As he knew deep down, even in his most depressed moments – and they must have been many – it was bound ultimately to triumph, even though for a time its eclipse might involve his country and the whole human race in misery and disaster. It was bound to triumph, because as he saw so clearly, it was that of the law of Nature, of the divine ordering of the world and universe. His championship of organic farming, of the family, of skilled craftsmanship and husbandry, did not arise, as his critics supposed, from sentimental or aesthetic motives, but from a strong and very English sense of reality which was increasingly reinforced by his study of the practice and history of good husbandry and craftsmanship. His accumulated knowledge of what practical Englishmen had done in the past, and a few were still doing in the present, to make the utmost of the natural resources of their country, was greater than that of any man living or, indeed, of any former scholar of whom I can think. 'Of the green mansion we know as England,' he wrote, 'they with Nature were the builders.' That the man who made himself their historian sometimes made mistakes and errors of detail was evidence not of want of scholarship but of the immense range he covered. It embraced almost every art of rustic life practised in this country during the past four centuries. That the greater part of that knowledge had been orally transmitted by its practitioners and seldom committed to paper made Massingham's achievement as a scholar all the more remarkable. That it was almost completely unnoticed in academic circles, or, so far as it was noticed at all, derided, is a measure of the degree to which the study of history in this country has become abstract and divorced from reality.

For Massingham, as I have written, was above all things a realist. He had his hero Cobbett's contempt for everything that was not orderly, useful, economical, and wholly efficient: not efficient, that is, merely in the financier's or monoculturist's sense, which, as he saw, was always in the long run grossly inefficient since it neglected the whole for the part, but efficient in the sense of achieving that germinating unity of man's skill and love with Nature's law – and, he would have added, of God's will – which is the source of all earthly creation. His eyes, as he wrote of Cobbett's, 'never separated what was useful from what was beautiful. . . . A smiling land and a smiling people living on it and by it, this was his earthly paradise.' He hated, like Cobbett, what he once more described during a war-time journey across Wiltshire as

> more gates that could not be shut than could, more gates broken than whole, tousled heaps of straw by the stacks, dishevelled combined fields, stacks growing out, tumbled or gaping drystone walls, ivy-covered trees, indifferently-ploughed fields, weedy pastures, dilapidated farm buildings, even barbed wire sagging or twisted.

The sight of the contrary, of good order, good husbandry, fine craftsmanship, wise and gracious living – the fruits of sustained industry, love, patience, faith, and balanced thinking – aroused in him an intense joy and thankfulness. No man ever loved his country more passionately or understood better all that was greatest in her achievement. From it he drew his courage and his inspiration:

> We came in sight of Romsey Abbey and both fell silent. That lion of lovely strength on its low mound was a rebuke to full despair, and we drew upon its healing power. What man did once he can do again, if only the conditions can be righted.[1]

Massingham's philosophy – of the indispensable link between God, man, and Nature – was like Charles I's head for Mr Dick; it crept into everything he wrote – on topography, horticulture, archaeology, history, literature, agriculture, music, painting, craftsmanship. Frequently it did so to his personal loss and detriment since it wearied the thoughtless or escapist reader of his otherwise charming writings and aroused the anger of the vested interests against which he unceasingly fought. Yet what a true, noble, and profound philosophy it was! How much deeper, for instance, even than that of such a greater contemporary thinker as Bertrand Russell or, on a far lower level, than that of that magnificent writer and brilliant but facile popularizer, H. G. Wells. The gods of our age and of the age that led to our disastrous age were false gods. John Massingham, increasingly throughout his life, fought for true gods. During much of his time he seemed to be fighting for them almost alone. When posterity surveys our epoch and its folly, this prophet who warned us of an impending fate will not be forgotten. In his last book, published shortly before his death, he wrote of a former forest on the Welsh border:

> All has now been so savagely cut over that hardly a tree is to be seen except the conifers of the Forestry Commission, themselves to be clear-felled. . . . Everywhere I saw the voids where trees once stood, filled in with a litter of bramble, willowherb and tussocky whitegrass. . . . The whole area is a dismal derelict waste, an upland hell and the bleakest of monuments to man's suicidal folly and cupidity. Wentwood as it was and as it is is a speck of that vast area of the earth's surface where the fire-roar and the saw-crash of the forests reverberate through the continents, Africa, Asia, the Americas, Australia, and now in our own country. Behind the prostrate trunks come the winds and the waters, scooping up and flinging into the seas and silted rivers the top-soil without which man is blotted from the world and will vanish for ever in the dust-storms of his own making.[2]

Whether we agree or not, we are listening to the counterpart of an Old Testament prophet. History teaches that, however unpalatable such warnings, it is as well sometimes to listen – and to heed.

Sir Arthur Bryant

Notes

1. *Wisdom of the Fields,* Collins, pp. 241-2.
2. *The Southern Marches,* Hale, p. 352.

ACKNOWLEDGEMENTS

Acknowledgements are due to the following for permission to quote from copyright material: to Mrs Penelope Massingham and to The Society of Authors as the literary representatives of the estate of H. J. Massingham for passages from *Remembrance, Shepherd's Country, Downland Man, English Downland, Chiltern Country, The English Countryman, Men of Earth,* and *Faith of a Fieldsman;* to Chapman & Hall for passages from *Genius of England, A Countryman's Journal, The Sweet of the Year, Fall of the Year, Shepherd's Country,* and *Men of Earth;* to Collins Ltd for passages from *This Plot of Earth, The Wisdom of the Fields, Where Man Belongs, An Englishman's Year,* and *The Curious Traveller;* to Dent Ltd for a passage from *The Natural Order;* to Robert Hale Ltd for a passage from *Southern Marches;* to Thames & Hudson Ltd for a passage from *Prophecy of Famine;* to Jonathan Cape for passages from *The Heritage of Man* and *Downland Man;* to Batsford Ltd for passages from *Remembrance, The English Countryside, English Downland, Cotswold Country, Chiltern Country* and *English Countryman;* to A. & C. Black Ltd as successors to T. Fisher Unwin for a passage from *Untrodden Ways;* to Bodley Head Ltd as the successors to Cobden-Sanderson Ltd for passages from *Through the Wilderness, Wold Without End,* and *Country;* Methuen Ltd for a passage from *In Praise of England* and finally to Lady Bryant for Sir Arthur Bryant's tribute to H. J. Massingham.

Personal thanks to Mrs Penelope Massingham for all her help, encouragement, and hospitality; to John Michell for his interest, help, and Foreword; to Liz Saxby and Satish Kumar at Green Books for taking on the anthology and to Richard Baxter for the use of a photocopier.

E.A.

INTRODUCTION

'In our depths we are a country, not an urban, people.'

1988 sees the centenary of the birth of H. J. Massingham, the natural successor to the tradition of English rural writers that includes Gilbert White, William Cobbett, Richard Jefferies, Edward Thomas, and W. H. Hudson. However, Massingham, through the great breadth of his knowledge, extended this tradition to encompass all facets of English rural life from pre-history to the twentieth century. His interests in literature, art, ornithology, archaeology, anthropology, geology, topography, agricultural and rural history were all brought to bear on his subject of England and its countryside. This wide perspective gave him an unrivalled position to chart the rural history of England and gives much weight to his opinions and conclusions. His views at the time when he was writing were unfashionable and very much out of step with the current thinking and, although aware that he might be regarded as somewhat of a crank, he persevered and today the values that he propounded are now widely accepted.

H. J. Massingham was born in 1888, the eldest son of H. W. Massingham, the famous radical journalist and editor. His father had left Norwich in 1883 to become the London correspondent of *The Eastern Daily Press,* a paper which H. J. Massingham's grandfather, a prominent Norfolk non-conformist preacher and radical, had helped to found. This family streak of non-conformity and radicalism never entirely deserted Massingham, even though he speaks of his conversion from an urban, liberal free-thinker to 'a Conservatism of a much older tradition'. H. W. Massingham eventually went on to edit *The Nation* and establish it as the leading Liberal political weekly from 1907 to 1923.

Massingham's early life was spent in London where he received the conventional middle-class education at Westminster School. This was followed by Queen's College, Oxford, at which first he read history, later transferring to English literature. On reflection he always felt that his education had left him ill-prepared for real life, certainly for the life that he finally chose to live. The one benefit which he did gain was his great knowledge and love of English poetry, especially the metaphysical poets of the seventeenth century, and this was to prove to be the first cornerstone of his useful learning. The second was his interest in bird-watching which came about when he spent a year convalescing in the country from peritonitis, an illness that prevented him from sitting his finals at Oxford.

When he had recovered, he returned to London to follow his father's career in journalism. With all the advantages of being the son of H. W. Massingham, whose friends included all the leading literary and political figures of the day such as George Bernard Shaw and Asquith, Massingham found it easy enough to get on, working for various newspapers and journals. He also contributed to *The New Age,* edited by A. R. Orage, a non-political magazine that stood for 'Guild Socialism', a theme closely related to the craftsmanship that Massingham was later to champion. It also introduced him to fellow contributors, poets W. H. Davies and Ralph Hodgson, and sculptors Jacob Epstein and Gaudier-Brjeska.

Massingham continued with this way of life up to the outbreak of the First World War, when a combination of his earlier illness and some sort of mental breakdown prevented his call-up. The early war years were spent in the Reading Room of the British Museum, and away from London in Somerset at Coleridge's Nether Stowey cottage and in Dorset in the Vale of Marshwood. This period of his life, which also included an unhappy first marriage, seems to be one which he preferred to forget as it occupies little space in his autobiography *Remembrance* and there is no mention at all of his wife.

He returned to London in 1916 to resume his journalistic career, this time working with his father on *The Nation.* Here he contributed various natural history pieces and helped edit the literary pages in which he was amongst the first to publish and review the poems of Aldous Huxley and Edmund Blunden who was to become a lifelong friend. But it was Massingham's involvement with one of the first successful conservation movements in 1919 that changed the whole course of his life and brought about his meeting with W. H. Hudson.

Inspired by one of Ralph Hodgson's poems he helped found The Plumage Group to campaign for banning the use of wild birds' feathers in the millinery trade. His commitment to the cause was such that he even got himself arrested for accosting passing fashionable ladies in Piccadilly. A visit to Queen Mary at Buckingham Palace was arranged through his father's contacts to enlist her support, but it was the meeting with W. H. Hudson for the same reason that was to prove the most influential to Massingham, 'Hudson laid hands on me', as he was later to recall, and from then on his life was to have a new direction and new purpose.

W. H. Hudson (1841-1922) was at that time England's greatest naturalist and ornithologist, but that pre-eminence had been hard won. He arrived in England in 1875 from the Argentine and for years he had struggled to make his living as a writer. Finally during the early years of the century his books on rural England, especially *A Shepherd's Life,* had established his reputation. Although their friendship only lasted four years until Hudson's death, Massingham was completely won over by the patriarchal figure of the great naturalist, and Hudson's influence pervades the rest of his life. It was through Hudson that Massingham began to question 'modernism', the Victorian cult of progress, and determined to turn his back on urban life. Hudson was to prove the catalyst that brought Massingham's underlying dissatisfaction with his early, literary London life out into the open and that made him resolve to embark on the exploration of 'the wild places'. It is ironic that Hudson, 'one of the last primitives and the greatest of countrymen', lived out most of life in England in London after his early years on the Argentine pampas and that his disciple, Massingham, followed the exact opposite course.

Massingham did not leave the city immediately, but 'for more than a decade of the post-war years my life oscillated between London and the country, and so reflected to a nicety the conflict in my own mind'. Another link with London was broken with his father's dismissal from *The Nation* and death soon after. However, it was during this period that Massingham produced his first country books that owed much to Hudson in content and spirit with titles such as *Some Birds of the Countryside, Sanctuaries for Birds,* and *Untrodden Ways.* It was not until 1926 with publication of *Downland Man* that Massingham's own voice really began to be heard.

Downland Man and its successor *The Heritage of Man* came out of his association with the anthropology department of University College, London. A visit to Maiden Castle near Dorchester, the greatest hill-fort in southern England, prompted Massingham to write to Professor Perry, head of the department, about the links between it and similar settlements in the Mediterranean. This resulted in him being employed as a researcher and assistant to Professors Perry and Elliot Smith of the anthropology department and allowed him free rein to explore the forts and barrows that litter the southern English downland, 'the churches and sepulchres of prehistoric England'. From all his work he became convinced of the link between prehistoric English civilizations and those in Crete and Egypt. It also introduced him in intimate detail to an area of England, Wessex, to which he always returned and which he always loved, and it added another layer of his understanding of the country. It was the second turning point in his life and 'planted within me the sense of continuity in the human panorama'.

The end of the 1920s found Massingham back in London in the world of literature, writing and editing various biographies, but in 1930 he was at last able to fulfil his promise to himself when he went to live for nearly two years at Blockley near Chipping Campden in the Cotswolds. The choice of the Cotswolds was really a matter of a chance, one reason being that it was near his sister, an actress who was appearing at Stratford. But for Massingham it was a fortuitous choice as he came to love the Cotswolds as much as the chalk downland. It was here also that the idea of English regionalism became firmly fixed in his mind, a regional integrity which in the Cotswolds was determined by the underlying geology, the Cotswold limestone.

One last commission brought him back to London before he finally settled once and for all in the country. Friendship with his then publisher, Richard Cobden-Sanderson, introduced him to the village of Long Crendon, near Thame on the edge of the Chilterns. Here, now married again, this time very happily, he had a house built which was to remain home to him for the rest of his life. His first attempts to be accepted as a countryman were rudely rebuffed by the locals and made him aware that as a newcomer from the city he had no prescriptive rights to be part of the community. So he set about re-educating himself, 'learning things of fundamental importance to the life of the community'. It was always Massingham's great regret that, apart from his gardening, he had no manual skills, especially rural ones, and that for him his life as a writer was somewhat unbalanced: 'If I could have my life over again, it would be with plough-tails in my grasp for the morning, and pen for the evening, or writing and wooding.' He states in *Remembrance* that he was prouder of his involvement, and that only as editor, in the book *England and the Farmer* than in any other of his previous books.

Massingham was now becoming established as a country writer and devoted the rest of his life to the subject in its many manifestations. He was invited to contribute to the Batsford series on English regions and he chose the two areas he knew well: *English Downland* and *The Cotswolds* together with the area around his newly adopted home, *Chiltern Country,* were the result. Even today, when much of the landscape has irreversibly changed, these books are still indispensable guides to the particular regions with their rare insight and their view of an historically integrated landscape. Massingham eschewed the turnstile, the beauty-spot, and 'the picturesque', all products of an urban mentality, and presented portraits of a living countryside, not one dead in the past.

> The meaning of what I saw in the two Englands I travelled, ours and our forefathers', had not altogether escaped me. In my wanderings over the country, I could not but observe that our own culture, the latest and therefore the most advantaged in profiting from the example of the others, is the only one from B.C. onwards which has failed to enrich its mother-earth, whether under the soil by cultivation or above it by its buildings.

It was during his extensive travels to research these topographical books that Massingham became acquainted with more and more rural craftsmen, specialists in each region, the bodgers in the beech woods of the Chilterns, the basket weavers on the Somerset levels, and it was by way of making up for his lack of skills that he recorded in great detail their methods of working and their tools. They became for him lighthouses of integrity and continuity in the increasing bleakness of the landscape. He amassed a vast collection of these craftsmen's tools, 'bygones', for which he built a small museum, 'The Hermitage', in his garden and which formed the basis of his book *Country Relics.* He later gave his whole collection to Reading University where they became part of its 'Museum of English Rural Life'. Craftsmanship, 'the foundation of all civilizations in the past', was for Massingham 'manual literacy', a continuous tradition proudly handed down from generation to generation, murdered, in his view, by the mass production of the Industrial Revolution. It was by observing these craftsmen that he came to understand their non-predatory relationship with their surroundings, their co-operation with nature, in short the ecological balance that they had to maintain to ensure their continuing supply of raw material. These observations led in turn to his recognition that husbandry too was part of a similar tradition: 'Husbandry is craftsmanship, the sum of all craftmanship.' He also saw that agriculture was coming under the same economic pressure that was destroying the craftsmen and this stimulated his interest in farming and horticulture, the major subjects of his later books.

A serious accident in 1937 when he gashed his leg on a rusty trough resulted in several years of serious illness and ultimately in the amputation of a leg. This disability obviously curtailed his extensive travels but not his creative energies. There followed a series of books containing a distillation of all his accumulated knowledge and wisdom, including his autobiography *Remembrance* and *The English Countryman – A Study of the English Tradition.* He also became involved at this time with a group of like-minded people, amongst whom there were his friends Arthur Bryant and Adrian Bell, to form A Kinship in Husbandry and he acted as the group's chronicler and editor. Its members included practical farmers and countrymen who were putting into practice many of the solutions that Massingham had proposed and thought necessary for the salvation of the countryside. Eventually he recovered sufficiently and with the help of his wife and various friends he was able to resume his expeditions, now using as his points of reference the agricultural equivalents of the craftsmen, those farmers who continued to use organic methods.

It was during this period that Massingham came increasingly to identify with another of his predecessors, Willam Cobbett, author of *Rural Rides* and great champion of the rural poor of his day. It was a progression from pure observer to committed polemicist, from a Hudson to a Cobbett. As today when we can find Massingham's writing both topical and prophetic, so too he had found Cobbett's: 'His themes are actually more pertinent to the 1930s than they were to the 1820s, because what Cobbett saw in the grain, we see come to the ear.' They did share many of the same themes, the championship of craftsmen and cultivators, the mutual interdependence of town and country, and the idea of use-in-beauty, beauty-in-use: 'If a thing is right, it looks right.' This could be equally as well applied to the siting of a village as to a well-pruned orchard.

> Cobbett did link the organic way of life with eternal truths and he was besides the most English of Englishmen. He possessed most of our virtues and few of our faults. To heed his vision of England would be to return to ourselves.

Massingham's description of Cobbett is an equally apt one of himself.

In his last books Massingham's main preoccupation is with organic farming and husbandry including his small-scale efforts in his own garden, and his advocacy of a return to traditional methods can be seen not as a rural reaction based on a warped nostalgia but as a logical conclusion of a lifetime's study. Although not totally alone, he was certainly in the minority when he drew attention to the loss of the hedges, the dangers of chemicals, both fertilizers and pesticides, the importance of humus, the turning of farms into factories, and the adulteration of food. He identified the sources of future concern and proposed his ecological solution, both well ahead of his time.

If the final conclusion of his work was a return to traditional methods of husbandry, the final conclusion for Massingham himself was a return to the repository of traditional values, the church. This was the final conversion from free-thinking agnostic to committed Christian, but the church he decided to join was the Catholic Church because for him it was 'the original Christian body and the only universal one'. Having never been baptized as a child, Massingham was baptized by the local Catholic priest in Long Crendon, but never really became a practising Catholic, and in fact was buried at the Anglican church. The village was the bedrock of English rural society with the church at its centre, surrounded by farms, cottages, and fields. This for Massingham was the rural trinity of God, Man, Earth, the simple basis of his faith. He expressed this trinity in various different ways – food, folk, faith and wholeness, health, holiness – but the dominant feature was the religious one, reinforced by the knowledge that Christianity, like all other religions, had been born out of a rural society. For Massingham a return to nature meant a return to God.

Had Massingham lived to celebrate his hundredth birthday, what would he think of the state of the English countryside today? As an early advocate of guaranteed prices for farmers, he would have been gratified to see the renaissance of farming, begun during the war and carried on during the next three decades, but appalled by a system that played straight into the hands of the agri-chemical combines. Now when the talk is of taking land out of production to be given over to golf-courses and houses, belatedly it has been realized that a return to more organic farming would reduce these surpluses. It would also provide better quality food and benefit the countryside by producing the more varied landscape associated with mixed farming and by lessening our dependence on chemicals.

Massingham knew, as Cobbett did, that the town and the country are interdependent, not merely with the country as a playground and vast picnic site, but in a more fundamental way, two sides of the same coin, an alternative way of life, and the source of certain values that somehow get lost in urban living. But it is the urban mentality that dominates, not only in what the planners allow to be built as their idea of village houses, which more and more resemble suburban developments, but also in the way that city prices have totally distorted the local market in housing in many rural areas. Inner city deprivation was the problem, but now there are enquiries into rural deprivation. The countryside which we have inherited is the result of human interaction with nature over thousands of years and, as Massingham pointed out, it is only in this century, with all its technical advantages, that we have left the countryside a less beautiful place. Short-term gain without any longer view is ruining this country's greatest asset, its countryside: 'Live as though you were going to die tomorrow; farm as though you were going to live forever', was a saying that Massingham used and it sums up succinctly the attitude that we ought to have to our own surroundings.

Any idea of individual regional identity has long gone as the dead hand of conformity lays across the land. No consideration is given to the use of local stone or styles in any new buildings and if one was suddenly deposited in a new city centre or a small village estate of 'executive homes', one might be hard pressed to identify where one was. The only fortunate thing is that the decline in building standards and craftsmanship will probably mean that very few of these buildings will be habitable, let alone standing, in 100 years time.

Massingham certainly recorded the twilight of rural craftsmen, the craftsmen whose trade was intrinsically bound up with the land. Although craftsmen still exist, they exist in isolation as some quaint throwback to the rural past, their existence prolonged not by old family traditions but by individuals drawn to an ideal of excellence in an age of increasing mediocrity.

The concern for a healthy diet has now become a reality. Massingham, in his own lifetime, was visited by the 'brown rice and sandals brigade', visits which he abhorred, but he would no doubt have welcomed the realization that we were removing most of the goodness from our foods. But can the age of the microwave sustain the trend or is it merely fashionable, a passing fad based on a more wealthy society? The distribution of the food as pantechnicons thunder from supermarket to supermarket shows that the food processing industry is alive and well while the distribution of local produce in rural areas is virtually non-existent; a healthy diet is far more prevalent in the town than the country.

On balance, then, Massingham might well be cheered by the movement to more organic farming, food that is less 'de-natured', and the greater awareness of the ecological balance, but dismayed by the relentless urbanization of the countryside, the growing centralization of power, and the continual fight needed to put over the balanced and sane way forward. The rampant materialism and consumerism, products of an urban mind, threaten to overwhelm the countryside, but the underlying rural values remain, waiting to be resurrected. These, as Massingham wrote, are spiritual and religious values, in that all religions have sprung from rural communities, and in England's case were Christian values. The recent news that a brand-new village is planned with all the amenities, but without a church would seem to be the culmination of the progress that Massingham so derided.

The phrase 'A Mirror of England' was first used by Walter de la Mare to describe one of Massingham's predecessors, Edward Thomas, but it is, I feel, even more applicable to Massingham whose range of ideas was far greater than Thomas's and whose concern for

England and its countryside is still far from misplaced. If in some part Massingham's vision for the future of the countryside is flawed, we cannot reject the whole, because overall he has been thoroughly vindicated. His concerns have been magnified by the passage of time and there is no doubt that his position as one of the major prophets of ecology should now be properly recognized and his reputation as a major contributor to English rural literature reinstated. The final words of this introduction should be his:

> The real division is between rival philosophies of life. The one believes in exploiting natural resources, the other in conserving them; the one in centralized control and the other in regional self-government; the one in conquering and the other in co-operation with nature; the one in chemical and inorganic methods imitated from those of the urban factory and the other in biological and organic ones derived from the observation of nature as a whole; the one in man as a responsible agent with free will to choose between the good and the bad and the other as a unit of production directed from above by an elite of technologists and bureaucrats; the one in the divine creation both of man and nature and the other in man as self-sufficient in himself with nature merely as the means for extracting wealth for himself. For of the two philosophies thus opposed, the one is leading the world on the road to ruin and the other offers the only way out.

NOTES ON THE TEXT

The various themes that run through Massingham's books in general coincide with times during his life, so the selection for this anthology is arranged chronologically, with a few exceptions. These are passages from his autobiography *Remembrance,* published in 1942, which are used in Part 1 to describe Massingham's early career in his own words, extracts 9 and 13 in Part 2 and extract 4 in Part 3 which are thematically linked. There is a logical progression in Massingham's works with one interest leading naturally on to the next, to the final development of his rural philosophy.

Part 1 covers the years 1888-1929 and includes Massingham's main interests in Hudson and anthropology plus pieces about his early life; Part 2 (1930-9) covers topography, regionalism, and craftsmen and Part 3 (1940-52) ecology, organic farming, and his general summing-up. In all sections there are pieces that show his increasing concern with what he saw as the destruction of the English countryside.

PART ONE

1888-1929

Birth

THOUGH I was born (the eldest of six) within sound of Bow Bells, and my earlier years were, with the exception of school at Littlehampton, all spent in London, I come, but for a slight seasoning of Huguenot, of East Anglian stock on both sides of the family, persistent in Norfolk since Anglo-Saxon times. To this in the main I attribute my ultimate disengagement from urbanism, which I consider the most important event in my life. My father was an emigrant and so in blood a provincial, and in the end this inheritance saved me from the prison of the pavement. He came to London not primarily because he was a high-flier on the *Eastern Daily Press,* but because his family was ruined by the economic system that has reshaped England since the Industrial Revolution. Cause and effect: the reason why I spent my formative years in the Great Wen was Dutch William's action in selling the royal prerogative in the control of money to the banks. Accordingly, my family history is that of hundreds of thousands of English countrymen from the village craftsman or land-labourer to the estate owner who were compelled to leave a pauperised countryside in order to become business men or factory-hands or clerks or journalists or artists or officials or dole-men in the once capital of modern finance and commercialism. My destiny was decreed long before my father met my mother, not by God but by Mammon.

Remembrance (1942:4)

Holidays

SOME holidays I used to spend at Old Catton, a few miles from Norwich. Here, in a biggish private garden and farmstead combined, lived a great-uncle, a bachelor whose house, also large and Victorian in emphatic provincial style, was looked after by one of my mother's sisters who, years afterwards, married my father . . . What I thankfully retain – a grove of palms in this unproductive waste – is the old white weather-boarded mill at

Trowse, just outside Norwich. It belonged to or was leased by the simple, warm-hearted man who married one of my mother's sisters. There I spent enchanted days fishing for bream in the millpool, wandering the luxuriant water-meadows on the banks of the stream, watching the great wheel thrashing the waters into a boiling churning seething hissing tumult of foam, eddy and spray, a sight that made my heart thump with the cheer of it. I threaded the purlieus of the dim, crowded, crazy interior, shuddering through all its beams as though with a palsy. I sniffed up that unforgettable savour compounded of sacking, grain, meal, dust, rope, mildew, water-weed, sawdust, and rotting timbers, sweeter than all the gums of Araby and like no other in the world. Nose, ear, sight, and even mouth, all were filled to the brim, and I could never bear to revisit that ancient, shaking, thundering, teeth-grinding, beneficent giant now that he is silent and unmindful of his procreant fields. The giant is dead.

But perhaps his influence is not, even to this very day. There was no more perfect intermediary between elemental nature and the works and sustenance of man than the old country mill. It was the gateway from the winds and the waters and the fertile earth to the whole economy of the rural and regional civilisation that is gone. It converted the cornfield on the spot into loaves as the water in Cana was turned into wine at the feast. It was the focus of husbandry, the hyphen between production and consumption. Incidentally, it was associated with one of the most delicate of the old crafts, dressing the mill-stones with the mill-bill. The vast machinery of distribution that has taken its place and has torn apart and dismembered the due sequences between the gathering of earth's fruits and our daily bread, this is put to shame for its wastefulness and all its monkey-tricks of adulteration for profit's sake by the simple, large-hearted offices of the country mill. It took what earth gives with one hand, gave what man takes with the other. Beauty and service, ethics and utility were made one by it. There were rascally millers in the days of old, but none has questioned the honesty of the country mill. Something of its majestic worth, of its soundness and truth, may have passed into me, all unaware, when I delighted in it by the banks of the Yare, and may have taught me, as yet unheeding, my first lesson in those interlinkages between man and nature which were to preoccupy me in later years.

Remembrance (1942:12-14)

Education

IT is singular that during the first twenty-one years of my life I was never taught anything whatever (1) about the social and economic history of my own native land, (2) about its geology, topography and architecture. This means the nature and disposition of the rock-floor that has conditioned its soil, landscape and vegetation. It means, too, the characters and differentiations of that landscape and the successive responses which man has made to these factors in his buildings, sacred and secular, regional and national. (3) About its agricultural history, the foundation of all civilised communities. (4) About its wild fauna and flora and their ecological relations with their particular environments. (5) About the food we eat, where it comes from, how it is grown and what is done with it between producing and consuming it. (6) About its archaeology, that is to say what its peoples and its places were like anterior to the written records. (7) About how to use my hands with the single exception of being able to twist a cricket ball a certain way with my left one.

I learnt nothing, that is to say, about the fundamentals, biological and otherwise, of our common existence upon earth. In other words, my education was urban only, and in the long run that is an education in the inessentials. All that I subsequently learnt about these things was of my own seeking; in what I do know about them (which is little enough) I owe nothing to my pastors and masters. But I have never succeeded in filling the blank and rectifying the omission of No. 7, and that to my mind has been the greatest loss of all. It has not only given me an unbalanced life but has enormously handicapped my studies, prolonged over a number of years, of rural craftsmanship. This I have long come to regard as the secret of the good life, the means to happiness and the true resolution of the inextricable tangle into which our social, economic, and indirectly our political life, has tied itself. The craft of living, the craft of working, the craft of content, the craft of recreation, these should be the aim of all education, since by them alone, interpreted in their widest meanings, is it possible to achieve the satisfaction of a full life. Hand and brain, body and soul, the individual and the community, it is in the pursuit and practice of craftsmanship that they win their difficult unity. There lies the answer to the riddle of wholeness and the goal of its quest.

Remembrance (1942:25-6)

The *New Age*

THROUGH my association with the *New Age,* I came into contact with some remarkable people, for there is no doubt that the *New Age* in its heyday attracted some of the best brains in the country. And of these, the editor himself, Orage, was *primus inter pares,* a powerful mordant personality, if a somewhat frightening one, and of an intellect with a cutting-edge that went through pretensions like butter. But he was more than an acid, even a ferocious critic; he was a genuinely constructive thinker, and his creative example permeated his whole journal, even though it was always a little arid. Orage was an ascetic both in mind and by nature, and so not truly balanced. He was the Cassius of pre-war journalism and a thorn indeed in Caesar's side. He had the Latin, geometrical cast of mind, rather than the English-Celtic which prefers the curve.

The great advantage of the *New Age* was its freedom: it stood apart from or, as Charles Marriott would have said, "withdrawn" from all political partisanship, from the caucus equally from the combine, and its views were inflexibly disinterested. It clearly perceived that economic freedom was a prerequisite of political freedom. It spoke from a cold searching ironic detachment that was very effective and in the hands of its most able editor, was the instrument of unacknowledged but a steely power. It did not attack parties but the unseen forces that pulled the strings behind them, and it grasped with assurance the roots of all modern discontents – the financial system, the child of what mediaeval society, religious and secular both, condemned with one voice as mortal sin, the sin of usury. The sin of buying cheap to sell dear, of getting something for nothing, of the unjust price, of self-interest as the ruling motive of the social life, the sin of beggar-my-neighbour, the sin of money as the final standard of value and not the medium of exchange between producer and consumer, the sin of confusing ends with means, the sin of the money-changers in the Temple. The sin of usury, but not like the mediaeval sin, because the modern usurer (the banker) loses nothing whatever when he lends to the borrower; he creates money out of nothing, and by putting the whole community in his debt. He produces nothing and he risks nothing because he invents the money he lends at interest. All he really hands over is a cheque book. All he does is to make an entry in a ledger. But he puts millions into pawn.

The crusade of the *New Age* against the money-power was, though interpreted in modern terms, well and truly derived from scriptural and mediaeval sources. Towards the end of his life, Orage retired into a kind of monastery, and the derivation of his thought, laic only in practice and principle, was ultimately monastic in origin. It proceeded (whether he acknowledged it or not) from the "thou shalt not commit usury" of the old Church. The doctrine was broken by its own guardians, and through the breach in the wall of that great principle was to march a strange procession – an obese autocrat concerned with his conscience, the Ironsides after him, Dutch William who sold his purse to the Bank of England, the great lords of the Enclosures, the Victorian frock-coats headed by Adam Smith, Malthus, Cobden and Darwin, and bringing up the rear the metal Juggernaut whose fuel is human blood. When the *New Age* attacked the citadel of Filthy Lucre, it was with the modernised weapons of the old. But the criticism was not merely negative and destructive; the creative element was the advocacy of the principle of the Trade Guild, and this was the perfectly logical conclusion. That was why the *New Age* was essentially non-party – the Trades Unions missed the whole point of the Guild System which was simply to give the artisan an interest and security in his work, in other words to turn him into a craftsman. The Union thought only in terms of earning a wage; the Guild in terms of work and its just price. The accent was on the thing in itself and not on what you got out of it, on the end not the means. The good life was not in acquiring but in making; it was the good thing, not the good thing out of it that mattered.

Remembrance (1942:31-2)

The Plumage Bill

THE record is a very mixed one, but now at long last after thirty years of wandering about the wood in circles, I did begin to learn something. I began to find a sense of direction. I started bringing things together. As usual, it was circumstance and not personal volition that guided my steps. I was at this time a heart-to-mouth friend of Ralph Hodgson, and, if armoured with love and the desire of good talk, you penetrated into the Hades of his small room near Adelphi Terrace, past his Cerberus of a bull-terrier and through the fumes of his shag which he smoked all day long with the windows shut, you gathered much oracular wisdom. The dog was named "Mouster" after the Mousterian culture of Neanderthal Man, and it was worth braving Cerberus for such pre-classical associations. Hodgson was as possessive about his poems as a broody hen; after writing a poem, he would lock it away in a drawer, hidden from all eyes but his own and often forgotten like a squirrel's nut. He would not publish more than about a tenth of what he wrote, and then reluctantly. So it was a note indeed of special favour when one remarkable day he showed me a long, foxed, dog-eared manuscript poem he had written about the Plumage Trade. It fairly throbbed with rage and pity, and a swell of music like though more measured than the thunder of the guillemot chorus among the ledges and pinnacles of the rock-face. It captured and haunted me to such an extent that from that day forth I made up my mind that something must be done to put a stop to this abomination.

To that end I founded a little society which was called the Plumage Group. We got out a notepaper bristling with good publicity names and entered upon a campaign for a new Parliament Bill (there had been several already) to prohibit the plumage of wild birds for millinery purposes. That campaigning was hardly more than window-dressing. Behind a bit of influence and the big names there were only three or four of us crying out in public: "This trade is damnation." We were helped by other societies and by a general if sheepish feeling that something ought to be done about it. One of the longest walks I ever went was along the red corridors of Buckingham Palace, to see the Queen in order to enlist her sympathy. I see myself now, walking on and on and on like you do in a dream, and the farther I went over leagues of carpet, the more my courage and memory of what I had prepared myself to say oozed out of my shoes along it. I seemed to get thinner and thinner as I went on and on, until only a shadow arrived. The Queen was so nice about it that I think she must have noticed she was being addressed by an ex-man. We went down to the London Docks to see the enormous piles of birds' wings almost of every family of the Class Aves distributed throughout the globe, and heaped up in mountains there under the roofs of the warehouses. This foul spectacle injected us with just the right spleen, and finally after one failure we got our Bill. I found myself in the singular position of sitting on a Commission under the chairmanship of the cultivated Lord Crewe, partly to enforce the new Act and partly to consider applications for the admission of certain plumages. Across the table sat representatives of the traders whom for months past we had been pillorying for the knaves they undoubtedly were. We had not bothered ourselves much about the etiquette of chivalry:

Here be rules, I know but one –
To dash against mine enemy and win.

But this by the way: the point of the Plumage Bill campaign was the effect it had in binding together certain disjunct and disparate elements in my experience of life, and so in carrying me a step forward towards, not the achievement (I never achieved it), but the conception of wholeness, which is holiness . . . But a more important revelation was into the true meaning of what science and general opinion united in describing as "the conquest of nature". It would have been impossible for so vile a traffic, engaged in such wanton methods of raiding bird-colonies in the breeding season, and entailing a butchery on so vast a scale, exterminating many species, imperilling still more and they the most beautiful in form or coloration and for the sake of satisfying the merest whim of fashion created by self-interest profit-seekers, it would have been impossible for so hellish a commerce to have persisted as it did if the human relation to nature were not resting upon a false basis. And what was this basis? Clearly it was predatory and acquisitive only. Was it man's customary attitude to nature? Was man the instinctive matricide? Was his only thought of his maternal heritage to rob, exploit and spend to his own vulgar gain the natural riches for his body and his spirit and his knowledge lavished upon him? Or was this approach to nature the expression of some widespread mental disease, contrary to the nature of things and man's own health of mind, but induced and fostered by an economic system confined to a particular phase in the history of Western civilization? Was it, in other words, a birthright from nature, herself predacious, or was it the effect of historical causes and sequences?

Remembrance (1942:41-2)

Meeting with Hudson

AND my friendship with W. H. Hudson, which began just before or just after the end of the War and endured up to his death in 1922, both kept me in London where he lived except while wintering at Penzance, and increased my maladjustment to urban life. He was a powerful influence upon me, more so, I think, than that of any other living man I had met. It extended beyond our mutual interest in birds, a fire in me which he fanned, though, oddly enough, it was rarely if ever ventilated between us. We, the master and the apprentice in the workshop of wild nature, used mostly to talk about books and places. Hudson, as his published letters amply reveal, was as willing a slave of the lamp as I have always been myself. We talked about the Plumage Bill, of course (one of his major passions), but that was all. No, Hudson's appeal to me was to something vaster, grander, more indefinite. I am not really an admirer of Epstein's work, but I do think that his Rima in Hyde Park conveyed something of what I mean. I was on the Committee which chose it – it was the only possible choice. I do wish I could remember Lord Baldwin's face at the unveiling ceremony just after Cunninghame Graham's speech. I think he must have expected something pictorial after the manner of Mary Webb. The only trouble about the Hudson Memorial, as Arthur Bryant told me years afterwards, was that no birds ever came near it.

Consider one thing about him which is not clearly apprehended. He was actually born before the repeal of the Corn Laws which marked the turning point in our history between an England depending on herself, her agriculture and her previous traditions, and an England of trade, cheap foreign foods, progress and urbanisation. Hudson was almost an exact contemporary in years of the Rev. Francis Kilvert, whose Diary delights in a countryside almost as foreign and far away as Hudson's own La Plata and Patagonia. He appeared in our midst, that is to say, as a sort of apparition, tall, stooping, eagle-like, from a world and a culture that lay at the back of our own, pushed out of sight like a bundle of rags that had once been a dress. It is queer, when I come to think of it, that I have known three men like eagles: John Sargeant, W. H. Hudson, and Henry Tonks. Hudson walked into our front garden from out on the moors. No wonder that we did not recognise him and for many years let him almost starve.

But his spirit reached us, who thought only of the earth as the last world to be conquered, out of a more remote past than the pastoral Argentine and pre-industrial England. He was a kind of plenipotentiary from natural man. From first to last, he was the great primitive, utterly cut off from the urban community in the midst of which he came to live and partially so even from the agricultural one that preceded it. *A Shepherd's Life,* perhaps the best-loved to me now of all Hudson's books, is the only one he wrote which specifically accepts the English rural tradition as bound up with what he felt to be a yet wider and deeper heritage, that of nature herself. Even in this book, he says next to nothing about the husbandry of Caleb Bawcombe's village. The works of man in local correlation and balance with the manifold aspects of nature by means of crafts and husbandry, the sovereign remedy for our modern servitude to the machine – these were not his theme. In a sense, he was too wild thus to extend his field of observation. He loved the "incult" too religiously for that, and his earlier associations with the nomad rather than the peasant no doubt contributed to, no, not this restriction – you can hardly call a man who took all nature for his province as limited – but his stopping short of the cultivated patch. He represented the corncrake and the rabbit in the corn, not the ear and the reaper. He was a man who was the very voice and presence of wild nature in human form, more so than ever Rima was. There could not possibly have been a contrast more absolute than between the industrial civilization of modern Europe and the man who walked into it from the wilds.

Imagine the effect of this wonderful being upon me at that stage of my life! If I wanted to go back to the Bass Rock, I had only to take the bus to Bayswater. His indefinable English, trailing about his theme like the shoots and tendrils of Traveller's Joy, has ever been my delight, and to this day it comes as fresh and undefiled as a spring morning. And why is this? Because it is itself like nature, slow in pace and growth, rambling in structure but suddenly throwing out a cluster of flowers, brilliant in colour or of a cowslip sweetness. No more personal style exists, and so to know the person himself, crotchety as he was, formidable as he was, was the heart of nature, her *human* heart, endowed with human feeling and intellect.

Hudson sometimes talked nonsense and was in his old age inclined to be wilful, but he was our greatest naturalist. Partly because knowing, feeling and seeing were at one with him, but also because he was nearest to nature of any living man I have ever met.

This sounds trite, but really it was a miracle. He made a gigantic stride over six thousand years of civilized life. He arrived, and there was a highly cultivated man in an absurd high white collar, sitting in his dingy fusty room in St. Luke's Road, who had stepped all that long way. How ridiculous he made our theory of progress! You could not conceive anybody less progressive than Hudson, and yet in most things, the things that matter, he was far in advance of what we call "modernism". He was a perfect example of progressing backwards. He drew from the deepest well of all. As Chesterton said, progress is extending and beautifying the garden of our home, and Hudson's was the original home of us all.

The power that he received therefrom is, I think, the explanation of the prodigious feat in reanimating his earliest years a sick old man accomplished in *Far Away and Long Ago*, that masterpiece of "putting the clock back". The fount of renewal never failed him: he was perpetually reborn at its springs, so that the octogenarian, drinking at the eternal source, put off the "windowed raggedness" of age and was clothed anew in the boy. It was an integration that overcame the separating agents of space and time. If he left something out, Hudson was yet a supreme example to me of one who had achieved wholeness not only of age with youth and present with past, but of art with nature, mind with senses, and of knowledge with intuition. Through Beginnings he touched Ends, the mysteries of the Unseen, and his imagination took flight into its borderlands upon the wings of that primitive animism which he described as the projection of his being into nature. Thus, you never find in Hudson any discord between fantasy, observation and the telling of tales. He was mythopoeic as is the early mind of man by nature, but it was by the use of that quality that he became of the kin of Traherne and Christoper Smart.

Another aspect of Hudson's wholeness is perhaps the most momentous of all, and one that has been missed by an age whose methods and aims are both separatist. That is the essential compatibility in him between primitive and Christian. "Consider the lilies of the field" – Hudson's whole life's work was an interpretation of the mind of Christ in uttering words to us now emptied of all context. He considered the flowers of the field and the birds of the air; his aim was to reveal their "peculiar virtue and operation" in exactly the sense of "consideration". The fusion between the old man and the boy by which Hudson became the master of time was carried forward into the philosophic plane of reconciling Galilee with Eden. I do not pretend that I recognised this at the time, but I do now, and I have no doubt that it unconsciously influenced me.

Hudson laid hands upon me. He did more than give a meaning to my pilgrimages into wild places. Through him I was coming to question the whole fabric of "modernism", and that could only lead, as it did lead, to the abandonment of my London life.

Remembrance (1942:50-3)

W. H. Hudson

IT does not speak too well for our generation that it allowed one of the most romantic figures in the world, a man of adventure and a wild heart, and one, too, old in the service of beauty, to die with hardly a turn of the head. This man was a primitive in habit of mind, and yet so modern that he directed the evolution of human thought and revolutionised the relations between man and nature. He wrote a dewy prose and like far sheep-bells, or a bullfinch making colour-tunes among the red-cheeked apples, and his own people, who, except for a tiny minority, had looked upon him as a rather inhuman oddity who wrote well upon a very limited material, dismissed him with the *hic jacet* of a note upon the passing of a "well-known naturalist." His few friends mourned him, but the world of men, which his genius enriched and but little regarded, paid him in his going with a heedlessness equal to his own. And he was more romantic than the heroes of most novels, and left the world to us to see and hear, all ears and eyes.

Hudson used to say to me that he had only enough to live on when he was too old to enjoy it. For nearly the whole of his literary life in London he was as neglected a writer as Jefferies, and with much less excuse, since Hudson is a classic and Jefferies was an artist only on the occasions when he was not trying to be one. Contemptible as often are the world's valuations of the great men living in its midst, Hudson certainly went half-way to meet them. I went many times to his cheerless lodgings in the Sahara of Notting Hill, but he was as oblivious to their gloom and its aridity as to the whole of London. In build he was spare and very tall, and he had a face like an eagle in the Zoological Gardens – noble, melancholy, remote, as though his thought migrated far beyond the "great wen" out to the sea-like Argentine pampas, misted with the fleecy plumes of the pampa grass. His attitude to the crowded city, the literary *salon,* the political arena was hardly even a natural aversion. His mind ignored them, as the upgazing eagle his iron bars. Few men of letters had even met him, and by scientific or pseudo-scientific bodies like the British Ornithological Union he was in his turn ignored, and well they might ignore him, since it was largely his disdain, eloquence and vision which have banished their lifeless and life-taking pedagogy to Laputa.

Until you knew him more intimately, Hudson had an almost Miltonic severity; his dignity and reticence were such as men used to associate with the nobleman, and a man so full of character, so solitary and aloof and a kingdom to himself might well give the impression of being roughened by prejudice and hardly approachable. He did indeed take some knowing, some getting through a rugosity of crust, some navigating his numerous but agreeably salty prejudices. But once over the bar and his friend found a personality that was gracious and affectionate, if melancholy and a little lacking in humour. A chief beauty in Hudson's works is that he is in them. Equally were they in him, and Hudson the writer – as was to be expected by anybody who can see how completely he expressed himself and put himself into his art – was an image of Hudson the man. The friend discovered that Hudson was not only what he wrote, but how he wrote. He was of a humble nature, no unbending don; he told as many stories in his talk as he does in his books, and his passion for and knowledge of English letters (a close student of his writing will guess how book-learned was this lover of the wild) were little behind what he knew about birds and flowers.

But what I remember most of my few years' friendship with him were his tenderness and generosity of heart. He was no more capable of pettiness nor any of the more vulgar emotions of humanity than an eagle is of twittering. Hudson had locked up in him, and the more fragrant for its secretion, a grace of loving-kindness, and when I think of him, and of his goodness to me, who had no claims upon him and nothing to give him, I think not only of the eagle but the linnet.

Nevertheless the impression of oddity and remoteness has persisted up to and beyond his death, and many things in him lend colour to it. His aloofness was immensely fortified by the conditions of his boyhood. He went to no public school nor university, and lived far away from the artificial restraints of city life, on the sparsely inhabited Argentine plain, as apart from all mental communion and social complexities as a human being well can be. He came to England in manhood, not to hear human speech, but the songs of birds, and to go, not to the Little Theatre, but the green-room of nature, avoiding contacts with civilised life as instinctively as wheatear and sundew. More than that, he persists in calling himself a primitive – "as inhuman and uncivilised as I generally am and wish to be," as he says in one of his greatest books, *Nature in Downland.* Nobody except Hudson could have written this passage in *Hampshire Days:*

> I began to grow more and more attracted by the thought of resting on so blessed a spot. To have always about me that wildness which I best loved – the rude incult heath, the beautiful desolation; to have harsh furze and ling and bramble and bracken to grow on me, and only wild creatures for visitors and company. The little stonechat, the trembling meadow-pipit, the excited whitethroat to sing to me in summer; the deep-burrowing rabbit to bring down his warmth and familiar smell among my bones; the heat-loving adder, rich in colour, to find when summer is gone a dry safe shelter and hibernaculum in my empty skull.

So desirable does this seem to him that he has a vision of the strange wild people who had lain dead for many centuries beneath the barrow, and they appeared before him in multitudes, all with their faces turned towards modern human life and habitations. And it appeared to the silent watcher in their midst that their "dark, pale, furious faces" expressed hatred and fear of the little busy, eager people who lived in brick houses all their lives and cared nothing for wild places and the sacred dead. He was wrong, for the town-dweller has a truer appreciation of nature than any nature-worshipping Neolithic villager.

This loathing of the tame and domesticated was a very pungent and genuine passion with Hudson. It led him now and then, not exactly to a contempt for human art, but to a strongly expressed conviction that at its highest it was inferior to the higher products of natural evolution . . .

His melancholy, both as man and, in spite of his gospel of the gladness of all sentient, non-human life, perceptively as a writer, strengthened the idea men had of his reserve and withdrawal. Mr. Clutton Brock, in an obituary notice, suggested that: "Nature was to him something which offered always what she could not give . . . and all people and things shared the beauty and the want in nature. . . . Most readers like a sadness which is resolved by a happy ending, a question put and answered, however foolishly; but Hudson's sadness could not be resolved, and his question was never put. Rather it was implied in the very cast of his mind, and gave both beauty and sadness to all he wrote." But this is the sadness of all men of genius with the metaphysic eye upon the phenomena of life, the stranger wandered from a reality of which this world is at once "the revelation and the shroud," and Hudson was not peculiar in possessing the common birth-mark of his fellows in soul.

His sadness had a more definite origin. He lived not only with his own, but a racial past, and this was the main reason for it. He felt, too, that the spirit of earth was mortally wounded and that the doom of her teeming birth was irrevocable. He felt this so profoundly that he was a sadder man even than Thomas Hardy is reputed to be, and it was strange that he never seemed to realise that he was the virtual parent and inspiration of a new driving force in the world, a revolution in values, a new and magical union of knowledge with imagination which will one day be a greater gain both to man and to nature than what has vanished has been lost. Hudson's example, his semi-religious veneration for life in itself, however manifested, is a deathblow to the predatory view of it, and a man who had begun a work of such grandeur might well go to his grave rejoicing. But he went forward with backward-looking eyes, and towards the end of his life he was a "too quick despairer" about the causes he most loved and had championed in the past.

And it is true that we have long passed that simple, that heroic and naïve period when it was the custom to rejoice at the "Conquest of Nature," and marvel at the heights and depths of our prowess as a species. That was an Age of Faith; here is one of question, of doubt, of despondency, of bewildered wandering and wondering among the thickets of the mind. Having added up the sum of achievement, we begin to ask what is its value, rather in the mood of a prosperous German possessing a hundred thousand marks. And Hudson's melancholy was a clear, deep reflection of this new spirit.

Untrodden Ways (1923:11-19)

Anthropology

IN one of my wanderings through Wessex, I came for the first time to walk the ramparts of Mai Dun, the greatest Iron Age fortified town in the world, two miles out of Durnovaria of the legions. I wrote to W.J. Perry of University College, London, about it, because I had read one of his books concerning the migrations of early culture from a common centre, and it seemed to me that there were interesting analogies between the hilltop settlements of our Celts and those of the pre-Hellenes about the Mediterranean.

To cut a long story short, our communications ended in my joining the anthropological staff at University College, then in the hands of Professor Elliot Smith and Perry, with a kind of roving commission to prospect the upland homes of prehistoric man in England and to act as their assistant both in research work and in preparing and editing their books and papers for publication. Perhaps it would be more accurate to say that I was enlisted as a light skirmisher in the then Battle of the Books raging between the Tylorians, or the advocates of the spontaneous birth of independent cultures with similar characteristics in various parts of the world, and the Diffusionists, who regarded civilisation as a complex derivative process originating from a single parent stock, favoured in its own germination by unique geographical conditions. This stock was Egypt, and these conditions were the regularity of the Nile flood in depositing its fertilising silt upon the predynastic wheat and barley lands, conditions repeated in no other quarter of the world.

It appears an eccentric move on my part that I should thus suddenly throw up my free-lance journalism for which I was now to have very little time, my book-writing about English birds and English places, and, most inconsequently of all, my hauntings of bird-animated wastes, in order to dip a studious nose into the customs and monuments of long-forgotten peoples. Actually, my course was not so irrational as it looks. The connecting link was wild England, since the antiquities of our pre-Roman forebears had, as it were, forced themselves upon my notice in those very places of loneliness and undisfigured beauty which I best loved to frequent.

Nor could I but observe how harmonious were their works with those of nature. The tumulus mounded the heath; the earthwork crowned the bluff or spur of the Downs in linear correspondence with the flowing and proportioned curves of the chalk or limestone range; the remote stone circle had once been a nucleus of human assembly with wild nature as its overseer and the object of its veneration. There had been a time when nature and man had not been at odds, when the high places of England and her elemental shrines had likewise been the sanctuaries of men, the founders of the England yet to be. This could not fail to arouse what of imagination I had, and more and more I had come to pry into the meaning of these memorials of ancient man, so massive, so perfectly blended with the natural scene, and yet not natural but the labour of peoples possessing the arts and crafts and husbandry of civilised society. They had left a mark upon nature more beautiful and yet stranger than she had been before.

I became so absorbed with the sense of this antique communion between human life and the wilds, that before had been tenanted for me only by the winged graces of the wild, that I flung myself into a vast new study that might well have daunted a wiser man. When, therefore, I came into contact with the writings of Elliot Smith and Perry, it seemed to me that I had found the key to unlock the magic door into the past. The secret was in the East, away among the pyramids uplifted from the desert like Silbury above the green waves of the Marlborough Downs. Once more the question was addressed to the incommunicative Sphinx, and I seemed to catch the faintest whisper, echo of an echo, in answer. But what I did not realise at the time was that my journeys among the churches and sepulchres of prehistoric England were intimately bound up with my journeys through the streets of London.

Remembrance (1942:54-5)

Maiden Castle

IT'S a queer thing: everybody in England has heard of the Taj Mahal, Rheims Cathedral, the sky-scrapers of New York, the Great Wall of China, and nobody at all outside the readers of archaeological works and county histories seems to have heard of Maiden Castle, the Mai Dun of the Celtic Durotriges and the Dunium of Ptolemy the topographer. If one had ten long lives, they would be too short for the discovery of one's own country, a love's labour but half won, and yet a year ago, I who am more than half-way through my little life, might have trudged on through the rest of it without giving a thought to Maiden Castle, a wonder of the world at my English door. It is the greatest earthwork that ever was, and in its company I spent some of the happiest and most tranquil moments of my life.

It is worth while to dwell upon the positive emotion of repose, the sense of security and of adaptation to one's surroundings that penetrate one's being in these silent, ancient places where a dead people or peoples, to whom the Greeks and Romans of one's schooldays are as children on the knee of time, have left large signatures and little else that once they lived. I have left all unrest and doubts and apprehensions behind whenever I have set foot upon or within these memorials, at the stone circles and avenue of Stanton Drew among the orchards of Mendip, among the dolmens of the Cornish moors, on the Norfolk brecks, and at Pilsden Pen, Bulbarrow, Dolebury, Camelot, Cissbury and other of the mighty camps. Surely, if one is at all sensitive to the peculiar effluence of places dyed in human associations, it might be just the other way round. The number of the dead exceedeth all that shall live, and as the curious modern mind uncovers layer beneath layer of human antiquity, sees the horizon of beginning receding and receding as it moves through continents of time, and surprises the primitive not as the infant but as the senile age of the civilized that went before it, the lives of one's self, of those who are dear to it, of all who live within this present inch of time, must appear as fugitive and of no more account than the shadows of the rooks on the Downland turf.

Perhaps, then, it is a sense of kinship with the earth, disencumbered from any thought of man, that brings so rare a serenity of soul? Ours is a valley civilization, that of our remote ancestors was of the hills, themselves remote from the idle business of the plains, while the earthworks are not merely spacious in their own designs, but command a greater spaciousness in which to roam in freedom and at rest. In their construction, too, the thought of the ancients went so far hand in hand with nature that they have become inseparables. These works of man have passed into nature, absorb hers into theirs, whereas with ruined castles and the like, her beauty has been gained at man's expense. She is the workman, but in the earthwork it is man who has made her more beautiful than she could have been without him. Ancient man has done his work and gone, but left nature, more nobly planned, as the memorial of his deathless labour. Love her, he speaks in the motion of the grass among the fosses and the ramparts, and forget me who live in her only.

In Praise of England (1924:147-9)

Yarnbury

YARNBURY next for another reason. When I first saw it, it was being scaled by the attacking forces spread out into units in irregular storm-formation, presumably to avoid those so formidable processes of the defence (enfilading and the like) upon which Generalissimo Allcroft gives us such copious information. They proved, however, to be little juniper bushes and they and I smiled slyly at one another, as I proceeded, single-handed (since the other hand held that trusty falchion, my umbrella), to breast the battlements.

Yarnbury, situated in one of the very loneliest and remotest parts of the Downs, lies north of the high road from Warminster to Salisbury and is connected by an ancient trackway with Old Sarum. Some seven or eight miles out of Warminster going east, this road joins the Amesbury road to the left at the little village of Deptford. A mile or so along this road is the camp with its three valla often rising fifty feet above its two and in places three fosses, so that it is fully on the scale of the Avebury earthwork.

Certainly the ancients had a marvellous eye for landscape, and my advice to a patriot in the undebased sense of the term would be to make a pilgrimage from earthwork to earthwork, and barrow to barrow, and use his eyes from their tops. He would be seeing his own home in all its delicate, lovable, generous and ever-changing beauty, as no guide-book could ever instruct him. This special faculty for selecting a site which both reveals and gathers up the true values of a landscape is well marked at Yarnbury. From the Amesbury road you just see the country round; a stroll over to Yarnbury, a hundred yards or so away and only a few feet if any above the level of the road and Stendhal's process of crystallization has mysteriously taken place. You have seen the Wiltshire plateau in all its intimate form and pressure, in, so to speak an image of its real self, once and for all. There are the Chitternes and Breakheart Hill to the north-west; the Codfords to the west; the Langfords in the Wylye Valley to the south; and the Woodfords and the Durnfords between Salisbury and Stonehenge to the east. Not that you see them, the dears, any more than you see 'The Voyage to Cythera' which the pairing of villages on this so finely moulded plateau irresistibly brings to your mind. You don't see them – this is just map-talk – for the demure little villages of Western Wilts all hide themselves within small cumulus clouds of trees. Down went my prejudice against clumps and groves of trees among the Downs, for nothing could be more delightful than the way they catch the vision as it roams and hovers like a kestrel over those effortless slopes. Nothing to break the congruity of the view except telegraph poles, and so Downland man had the better of it, for instead of them he saw droves of the great bustard, very possibly the only real difference he did see between our times and his. And in a perfect hoop of light, the horizon binds and blesses the composed and flowing and sun-dappled scene.

Downland Man (1926:235-7)

Trackways

All the green roads lead to Avebury as clearly as do our metal ones to London on an ordnance survey map of the home counties, and as inevitably as the watersheds that form the ribbing between the Upper Thames and the Severn, the Kennet and the little waterways of the south, are riveted upon the triangular plateau on which Avebury stands. It is becoming slowly apparent that all the vestigia left by the Downland occupants of pre-Roman Britain, earthworks, trackways, hut and stone circles, barrows and the scars of ancient mining works, are the ruins of one vast architectural whole, constructed upon a definite and systematic plan. It is not only that there exists, could we but grasp it, a topographical key to the unification of the highways. To possess the archives of this organized Transport Union would not be the end of the matter. The trackways join hands with the more ancient of the earthworks, the circles with the barrows, the barrows with the earthworks, the circles with the trackways, and each of them separately and all of them together with the mines. These are all the leaves, scattered, foxed, torn and barely decipherable, of a single volume, part of a set wrinkled deep in time, written in a foreign language, but very history. And when we have put the leaves together and then the volumes, and read them from first page to last, we shall know many things at last of which we now possess hardly a glimmer, and that knowledge is going to burst the safe and studious walls of the archaeological hermitage and throw its beams upon the world as it is to-day.

But I feel far from comfortable in the grandiose robes of the prophet, and hasten to step out of them and back to Avebury, seated on the plateau between the Pewsey Valley and the pastures of North Wilts, there in the centre of its web of trackways. With their chains of earthworks and barrows, like the knots and rugosities along the underground root-system of a leguminous plant, they and their Milky Way of daisies come trolling along from the Wash and from the Channel, from Salisbury Plain and Stonehenge to the south, from the North and South Downs with their network of shafted flint quarries at Cissbury, from the Chilterns and the Cotswolds with their long barrows, from the Purbeck Range and the Dorset Downs with their massive earthworks carved some to the contour and all to the measure of the hills, Maiden Castle, Ham Hill, Badbury Rings and the others; and from the Bristol Channel over Mendip ridged with earthworks of stone, pitted with the ancient workings of the lead mines and embossed with the stone circles and avenues of Stanton Drew.

It is impossible to grasp the meaning of Avebury, unless we fix it as the node of an intricate geometrical pattern, and that we can only do by gradually working outwards from the centre of gravity. To the east, the Ridgeway throws out a tributary to Inkpen Beacon, the pivot of the Hampshire Highlands, right to the centre of the great camp (Walbury) on its summit with its long barrow outside. The main line follows the chalk south-westward to the mouth of the Devonshire Axe. Thus Hampshire, Devonshire, Dorset and Wiltshire are made one. Another greenway passes various long barrows and camps and on through Stonehenge to Old Sarum, where it joins the Roman Road (built along the line of older and curlier trackways) from Winchester over Mendip top to the mouth of the Somerset Axe in the Bristol Channel.

A trackway connects Streatley and Hitchin in the Hertfordshire chalk region, and from Streatley the Great Ridgeway that connects Avebury with the Devonshire Axe crosses the Berkshire Downs to the holy city. The Thames, the Wiltshire Avon, the Wylye, the two Axes, the Stour, the Parret and the Severn are out of sight but not of mind of one another, for currents of thought pass from one to the other and many other rivers with them, and of all these interlacing threads of communication Avebury was the brain and nerve centre.

From the Winchester district, again, trackways run over the North Downs, and again south of them past Selborne and across the southern watershed, and south of that again to Butser Hill where the South Downs extend eastward for sixty miles to Beachy Head. A trackway crests them all the way. From Hitchen, the Icknield Way, which is the site of an ancient trackway and becomes the great Ridgeway in the middle of the Berkshire Downs, travels through Cambridgeshire on to Thetford, and itself passing on eastward, spouts out a turf jet northwards past the flint quarries of Grime's Graves and on to the Wash. Derbyshire is linked on to Wales and the Cotswolds by other trackways, South and North Wales had similar arteries of communication with the Mendips eastward via the Bristol Channel and south-eastward via Gloucestershire with Avebury, while other nerve-fibres kept Derbyshire in touch with the Yorkshire Wolds. Yorkshire, Westmoreland and the Cheviots were likewise threaded by trackways, and Lancashire, Herefordshire and Shropshire (where there are also 'prehistoric' remains), were not isolated from Wales and the Cotswolds. Such a complex of lines and settlements reminds one of a modern railway system with Avebury as Paddington, Euston, Victoria and Liverpool Street all fused into one junction.

Downland Man (1926:61-4)

England Laid Waste

THE face of England has been moulded by the social changes of English history, every wrinkle, every fold, almost its expression, as the human spirit and sometimes even the human face divine, by the vicissitudes of its experience of life. For aeons after the last retreat of the glaciers England as we know her remained in her wild babyhood – forest, desert and the bare summits of the higher hills, while palaeolithic man made no impression whatever upon her features. But with the coming of the megalith-builders, England received her first transformation at the hand of man. It is a remarkable tribute to the first civilization that laid the foundation stones of the English nation that it left English country actually more beautiful than it found her. On the chalk downs, the granite moors, the open heaths and limestone uplands, the large gestures of Iberian and Armenoid after him positively refashioned the contours of the hills into their earthworks, tombs and agricultural terraces. They added a new dignity to Nature without departing from the harmonious graces of her downland and so permanently kneaded them into continuity with human history. The literalism of their religious beliefs caused them to select for the shrines of their dead lords those sites which demanded the greatest stretch of landscape, with the consequence that we ourselves in a civilization which has dealt less scrupulously with the qualities of English scenery cannot secure an uninterrupted view of down, heath or hill-top occupied by the men of the megaliths without the eye being arrested by their labours in earth or stone.

Not only do these monuments take hold of and dominate the scene below them, but themselves catch in smaller compass the rhythmic curves and folds into which so serenely flow the lines of the downs. What would Stonehenge be without Salisbury Plain, and even more pertinently what would Salisbury Plain be without Stonehenge? All the visible landscape from Stonehenge belongs to the ancient temple of the sun, as the circumference to the hub of a wheel. A long barrow on the downs is a toy down in itself, while the great pyramidal hill of Silbury, near Avebury, has become Nature's but for a certain stamp of apartness in its form. Nor would it be true to say that these monuments, adding fuller lines to Nature, have gathered beauty because they are half as old as time. The centuries whose procession has passed over them have altered them little more than have the winds. Where they were once white they are now green – that is all.

The Celts, Romans and Saxons introduced much less significant changes in our landscape than the national vanity of our text-books will admit. The Celts had very little to say for themselves except Arms and the Man; both for burial and ceremony they mainly used the monuments they found here, while they undertook no fresh adventures into the forested plains. The Roman settlements were an exploitation rather than colonization, and the poor relics of their handiwork in comparison with that of the megalith-builders reveal their anxiety to get what they could out of the land rather than to enrich it. Many of their military roads were but a straightening out of the megalithic trackways, and here Nature had more work to do in softening their rigidity. Nor can straight roads, as the modern motor road so vilely demonstrates, ever become familiar with English landscape. It was the twisty, curly lanes and highways of the Middle Ages which made England more truly and variously herself. They, like the pre-historic greenways before them, established the principle of the winding corner that so exactly tallies with the shy, evanescent revelations of English country beauty. The share of the Saxons in maturing the configuration of the land has been grossly exaggerated through the political partisanship of Freeman and Froude. The Saxons were lowlanders who made small clearings by the sides of streams; but whether for weal or woe of landscape beauty, their small islanded homesteads can have made little more impression upon the length and breadth of English swamp and forest than has the rubber settlement in the Amazonian jungle to-day. Possibly their strip system of cultivation caused parts of the valleys to look like the striated vineyard patches in the southern plains of France. There is no comparison between these horizontal ribbons and the generous corn platforms of the megalith-builders.

The age of the tower and the keep was followed by that of the monastery and the cathedral. The first period accentuated that appearance of concentrating upon specific points of land which was characteristic of megalithic England, while the second dotted the wild with rich gardens. The cathedrals gathered townships about them and the abbey and the monastery set the style for the English country house with its park, lake and spinney abutting on the wilderness. The landscape gardening movement in the eighteenth century greatly artificialized the process. Such changes as the drainage of the fens, the clearance of the forests, the demarcation of common lands, the growth of cultivation extending ultimately (as in Cobbett's time) even to the tops of the downs, and the establishment of the yeomanry, proceeded in a continuous development from the Middle Ages onwards without being seriously or for long periods interrupted by the wars of the nobility, the Black Death, or the tragedies of social oppression which make our history only less bloody than its complementary prototypes. The face of England was changing, but so slowly that only great leaps of time would have marked the closer and closer intermingling of the eternal rhythm of Nature with the works of man.

A precipitous and revolutionary change was, however, dumped upon England by the enclosing of the common land in the nineteenth century. But the irregularity of freehold system, both in time and in extent, is not so well recognized. Hurstbourne in the Hampshire highlands, for instance, was for long years the king's land. From him it passed to the Tarrant Monastery in Dorset, and was then acquired by a family who leased it to a current profiteer. He it was who began extensive enclosures of a country whose varieties of ownership had hitherto been followed by no remoulding of its essential outlines. It might be asked why English country means principally hedges enclosing meadow land and French country still retains its open, boundless plains, though the process of the seizure of the common lands was the same in both. The answer is that the enclosures in England were inaugurated by rich men on a large scale, whereas the French Revolution consolidated the peasantry in the possession of their land. The enclosures bore far more heavily upon the richer yeomanry of England than upon the peasant proprietors of France.

Up to the industrial revolution, man's relations with Nature were in effect those of a co-operative partnership in a mutual interchange not perhaps of goodwill on either side but certainly of benefits. Even the wanton expropriation of the enclosure system did no harm to Mother Earth, only her labouring sons. But with the coming of the Railway Age, followed in furious sequence by the Motor and the Bungaloid Age of to-day, these bonds of communication were abruptly broken, and Englishmen began to impose themselves on the English countryside in the spirit of conquering aliens. The beginnings of the decline of agriculture corresponded with the pimpling of the north with industrial cities. Thistle and factory together combined to invade agricultural England in precisely the same way as Puritanism and promiscuity have made an alliance of opposites to destroy the conception of romantic love. The great wave of industrial development is already spending its force in the north and has begun to roll its sooty waters south, as industrialized towns like Reading and Oxford clearly warn us.

Other multifarious consequences of man's industrial mood of conquest over Nature have manifested themselves in all directions. The mongrel suburb has destroyed the particularity of division between town and country, in which each, living side by side, was true to itself. The oak, the beech, the lime and the elm, which shared their individualities between human tradition and natural nobility, are being replaced by the monotony of disciplined conifers, as standarized as the commercial mentality which has ordained them. The harsh lines of the quarry obliterate the green rollers of the downs. The motor road, inhuman, unnatural and altogether relentless, drives like a ram through the countryside with as much regard for its forms and design as a hot poker drawn over a carpet. Its great scars across the face of England lead us towards what Professor G. M. Trevelyan calls 'the mechanized landscape of the future.' The old roads, often buttressed with primrosed banks, and so truly modelled to the country qualities on either side of them, give way to these great tar tracks with their concrete borders, rows of equidistant trees, metal vomit of petrol stations and bellowing advertisements. The builder riots through the land like a skin disease, spilling scarlet hutments all over Salisbury Plain, making fungus pleasure towns sprout over the turf solitudes of the downs, putting up red brick in the stone countries. The old woods are grubbed up – Harewood Forest, the scene of W. H. Hudson's romance, *Dead Man's Plack,* disappeared during the War – and the starveling wire fence evolves in the march of civilization from the hedge with all its prodigalities of life, colour, form and line. It is a melancholy and ironic reflection that the wide distribution for the first time in history of a real love for the country should correspond with a period in which enlightened bodies like the National Trust have to wrest inches of untouched England from the devouring grasp of Progress.

Yet as the horde of speculators, company promoters, advertisement agents, country-house builders, bungaloiders, signposters, petrol-pumpers, river-polluters and all the motley caterers of profit and pleasure erupts in lava streams over the land, it should be possible to detect some method or guiding principle in the madness.

Let me take some particular examples of the way Englishmen are dealing with England and try to gather from them where or what is the heart of the malady. The street outside my house is planted with lime-trees. Every year, therefore, the servants of the Urban District Council assemble with the abhorrèd shears and proceed to cut down their twigging to the very bone. When the trees are leafless, they look like the next morning after a concentrated air-bombardment upon my particular street the night before. When a starveling leafage bashfully appears, these forlorn trees only lack plate-glass in front of them to resemble a shop-window exhibition of a more than usually nauseous decorated wallpaper. Since the last thing that would ever occur to my local government would be to plant wayfaring trees, hawthorns, rowans or other arborescent diminutives, it is local government itself which in this respect makes the Town Planning Act the dead letter Professor Abercrombie admits it to be. Or take the Forestry Commission. This worthy body is setting an example of beneficent industry by turning the unique Breckland of southwest Norfolk into a parade ground for conifers, equidistant, each the spit of its brother and all of them set out in standardized rows as though the voice of Nature had just bawled 'Attention!' The Commission is going to do the same for the New Forest. In its zeal for making profitable citizens of natural heaths and woodlands, this Government institution has an eye for pitching on the only two landscapes of their kind left over from the unkempt England of the past. And the Brighton Town Council that has made itself recently notorious – its Tories, Liberals and Labour men combined to plot part of the English downland near the Devil's Dyke under a dirt-track for bet-upon and bellowing motorcyclists. This unabashedly dissolute scheme for ravaging the downs, for making a profit out of them by pandering to mob excitements, was the plan of a local authority who no doubt take the utmost moral pride in defending Brighton from any breaches in decorum.

The Heritage of Man (1929:294-301)

PART TWO

1930-9

Regionalism

THE intercommunion between man and nature by way of the infinitely varied achievements of the craftsman – this I believe to be the key to the Cotswolds. The stone was the link between the human and the natural scene; it had inspired men and fashioned the hills. I was present at the dying flicker of the old culture which had survived the vicissitudes of four millenia. What had closed the door was the idolatry of material progress, but the door was just ajar to let me see into it.

All these things took time, years of time, for me to see, and I took still longer to grasp all that they meant. Ten years elapsed between my first seeing Campden and last seeing Winchcombe. I did not therefore jump to conclusions – is not haste the curse of the age, anti-natural to the core? – but they found their way into my being by a process of saturation, greatly helped during my two years at Campden by my constantly mixing with the villagers at the Baker's Arms and the Eight Bells, sampling their humour and learning their ways which have or had the large simplicity and generous lines of their country scene. Place and people became one.

Primarily, then, I became a convert to the idea of the region. Regionalism means all the factors of a given region, and each one is woven into the other. The parts, the units, the cells, are not separable, and the region itself is what binds them together into a living whole. The region is built up out of twin irreducible essentials, its geology and the soils evolved from it in correspondence with the native types of vegetation. These in turn condition the architecture, husbandry and crafts of the region, expressed in a variety of forms that extend over their general history and prehistory but are included within a single general framework – the land and the rock. Thus a specific quality manifests itself in the complete presentation of a region, in precisely the same way as it does in a work of art. A region thus presented *is* a work of art, achieving its self-sufficiency both by accepting and transcending the limitations of its material, as all such works do that are worthy of the name. The human contribution is as integral a part of the landscape as is that of its flora and fauna; man and rock are the end and the beginning of the masterpiece which is the region. The difference between him and all the other elements that interact with him is that he is able to contemplate them as the organic whole they are and were meant to be. Thus the part (namely, the place or region) is a whole in itself, and it is through that wholeness in the small that we measure the unity of the cosmos. It is impossible for us to measure wholeness at all except through the part, the particular, which inductively, by way of the historical method, leads on to the consideration of the general – the individual in the place, the place in the country, the country in the world, the world among worlds.

All this reads as the merest commonplace, and so it is, because it is fundamental. But it is a principle which the Industrial and Machine Ages have entirely reversed and so violated. Generality from the top, not particularity from the bottom, has been both their law and their action, as witnessed by mass-production, centralisation, cities like counties rather than towns, mechanical uniformity, imposition from without as opposed to creation from within. All are symptoms of a radical topsy-turvydom. The Cotswolds, on the other hand, offered me a perfect example of regionalism both in space and in time. The region showed an exuberant diversity within the confines of its oneness. It was pertinacious throughout not by imitating the past but by re-creating it and so making it, as it were, coterminous with the present and an unerring guide to the future.

Remembrance (1942:81-2)

Cotswold Limestone

ALMOST at last I am persuaded that this secret of Cotswold, and the bond of union between all its differences of natural and man-made particularities, emanates from the limestone itself, the bones of the land. It is unlike any other limestone in colour, in texture, in plasticity. Raw beneath the soil, it is of a most delicate buff colour, deepening in certain conditions to orange, but, after weathering and long exposure, tones to a warm soft silvery grey, of the lichen shade which mantles it. The transformation of the stone from buff to grey affects its surface, so that, sole among limestones, it possesses the quality of retaining the light and so of quickening and responding to the moods of light. Lastly, it is highly porous and, for all its durability, wonderfully amenable to the graving of the frost, the tools of all weathers, the erosion of watercourses and the chiselling of man. In Cotswold stone there is no dourness; it is as yielding in its stone fashion as the turf which rests upon its shoulders is resilient.

These qualities explain many things. I have never in all Cotswold landscape seen one hard or angular line. The rhythmic effects of hill and valley – so different in their several features and yet interchanged in such rich profusion, and flowing in and out of one another in an everchanging series of variations upon the main rounded theme – are primarily due to this adaptability of their foundations. In Gibbs's *A Cotswold Village,* the wolds are uniformly described as "bleak." This is just the kind of criticism one might expect of a man who probably never trod them except to kill wild beasts and birds, and whose park-like mentality shuddered at an open sheepwalk. I have walked the wolds in the hardest weather, in rain and snow and hail and through the trance of frost, but never sensed that aloofness and estranging desolation of the Cornish, Yorkshire and Derbyshire moors, of Mendip and Dartmoor, of the Welsh mountains, and even parts of the Wiltshire and Berkshire Downs. The lines of the wolds are too multiform in sweep and curve, too much diversified with woodland and grove, too well pocketed with waterways, hollows, vales and rifts where most of the villages lie snug, to earn such a title. The stone itself is not cold enough. Yet these wolds, as in all limestone country left alone, are well and truly wild, wild as their autumn skies, but it is a wildness that no more frowns upon human feeling than their limestone is chilly to the eye. On the contrary, they welcome the wanderer within their folds and expand rather than dwarf his thought.

Cotswold buildings – and where else in England are there so many beautiful buildings to the square mile? – carry the epic of the limestone into a new measure. I have never been able to see any break in the continuity between the masons of Rollright and Belas Knap and the masons of Campden and many a humbler chit of a village. Indeed, the dry-walling of the horns of Belas Knap is repeated in the stone walls of the sheepwalks and many a rustic bridge over the little rivers, until, as the centuries passed onwards, like cumulus after cumulus over that immense sky, the masonry grew rougher, so that the little bridge over the Windrush by Condicote cannot compare in skill of dry-walling with Belas Knap, more than three millennia older. And the roads – there is the continuity once more; the roads of golden russet overlaid upon, or passing into, or brightening out from, the grey or grass-covered track-ways over the wolds – Buckle Street and many another wagon-way and elf-way of loneliness, strewn with flowers.

On the high lands, one is continually in presence of the sky-world. Not only do the clouds stride over the land in their shadow-forms, but the configurations of the land meet the skies in cloud-like masses. In intercourse with these two planes along every foot of the way, one meets a third plane, some cot, barn, tower or farmhouse which, true to the nature of the stone dressed and carved in the traditional Cotswold style, appears to grow straight out of the soil in grace and accord with it. The stone expresses itself in a new idiom. "I will be a house," says the stone. That style and the impressionable character of the stone draw into a common brotherhood every type of building constructed from them. Barns grow towers and are called churches; cots expand and magnify themselves and are named manor houses. But gable, roof-tilt, dormer and set of the wall are the same, modified to suit different conditions. The simplicity of the barn is retained in the church, of the cottage in the manor house, and a fundamental unity of design governs them all. The true beauty of Cotswold houses lies as much in their democratic kinship as in their intrinsic harmonies of line and fitness to the soil. All this is implicit in the primordial element of the stone, like the lily in the bulb.

In an inspired way, the old builders have discovered the shapes that Cotswold stone best likes to assume, just as the waters, winds and vegetation have worked it into larger mouldings. The stone offers itself to become in masterly hands the tower of Coln St. Denys or a pigstye, a mighty ridge or tree-girt cup for winds to sleep in. But because of its willingness, its plasticity, its inborn virtues of hue and texture, the essential nature of the stone is never lost in all its divers manifestations, and in this elemental bond all beauties, whether formed of Cotswold man or nature, are united. When I have asked myself why the meadow crane's-bill between my fingers is larger-petalled and bluer than elsewhere, I have only been able to answer thus. It is the blessed nature of this Cotswold stone, "by some hid sense its Maker gave," which renders all that proceed from it, this flower no less than the slow undulatory speech of its rustics, all fair and all akin.

Thus, all things in Cotswold possess a unity whose primary source is the quality and nature of its limestone. So much for the matrix, but into what essence of beauty this unity is distilled, that escapes me. Partly I can disintegrate it into its several elements, its various rhythms and phrasings, but what it is I can answer no more than the philospher or the man of science can answer, "What is life?" I only know that it is present in the air I breathe upon that madder soil, and that even here in London I am haunted by its fragrance.

Wold Without End (1932:287-91)

Rollright

TO-DAY saw the first appearance of the south-west wind trumpeting over the uplands like a herd of elephants, but here in this shallow cup not even bending the tops of the trees. But there is a foretaste of cold, though there are still fields of blue sky between the thick hedges of the clouds. Next day it happened so, and a thin "whiffle" of snow on the wolds. It seemed to me just the day to pay a flying visit to the Rollright Stones, so I turned my wheel towards Chipping Norton and then left at the cross-roads along the ancient track that three and a half millennia and more ago saw men with packs from Wales and Wiltshire passing along it like an African safari. I had no light passage, for the road was filmed with a "gleer" of ice, and pockets of snow lay along the wayside grass. But I crunched and splashed and slid through it until that eerie circle of eight pines beckoned me along the ridge, and I halted in the lea of the stone circle as dusk was falling and a low moony sun was balanced on the edge of the western hills. Its heat and radiation were shut off from earth as though it shone through glass.

A day indeed for the "Stwuns," for they clash with smiling bee-busy summer in their remotely sombre gaze over purple distance. A round sixty of them there are, and now many of them were capped with snow in place of the wheaten cakes of traditional observance. In the failing light, with the earth contracting under my feet into stubborn frigidity and streaks of snow lying in odd places and the long smooth planes of the hills lifting and falling as in some cold dream through the haze, they looked their legends indeed. They looked, these animated stones, archaic priests and lords petrified, as though at any moment they might set forth, headed by the hawkheaded King Stone on the long barrow opposite, and stalk down the long slope to drink in the stream below. Now no man regardeth them, only we curious ones come to search out their stony bible, read in it the beginnings of so many of our woes, and yet observe how nature has honoured their works. It has covered over their follies and given them the dignity their fantasy once assumed was theirs. Nevertheless, something remains of that strenuous faith which practised its rigid cult of the dead upon these solitary high places and high above the panorama of winter I felt it as something hostile to the warmth of reality. "Drive your plough over the bones of the dead," was Blake's comment on the heritage of thought the megalith-builders have left us. So I left them, beautiful in their cold desolation, for my friendly cot. I was of them once, as we have all been our ancestors, and from the mortmain of the past all our lives we struggle to free ourselves.

There is no place in England more saturated in the archaic past than Rollright, and old Camden was right to call it the second of the seven wonders of Britain. As Evans – Minoan Evans – wrote of it in the sixth volume of *Folk-lore,* the pre-Domesday Book name, Rollright, Roeldrich or Rolldrich, is a corruption of Rollandriht and Rollandrice, the right (jurisdiction) and kingdom of Roland, Charlemagne's Roland who in mediaeval ballad ("King Lear") and folk-lore was a supernatural hero of popular epic usually associated with the Crusades. Roland's links with the megalithic ages were numerous, and, when heathendom was routed in Charlemagne's campaigns of the ninth and tenth centuries, the figure of Roland was erected as a substitute for the "stocks and stones" of the far older faith. Such was the process of the transition from paganism to Christianity. The gods were changed, but the forms and ritual of worship were largely maintained upon the old sacred sites. In Germany, Italy and Spain many a menhir and dolmen were called by the name of Roland, and became centres of kingly power in conformity with the megalithic obsession of divine kingship, and so of immortality.

There are many striking analogies between the Rolandsäule in Germany and the local legends of Rollright. As the King Stone comes down to drink in the brook of Little Rollright Spinney when he hears the clock strike twelve, so moved the statue of Roland from his pedestal. If you chip off pieces of the King Stone, the wheel of your wagon will lock, and, on the Continent, Roland represented the old Saxon god, Wedel, who held a wheel in his hands, both legends suggesting the solar disc. In English folk-lore, "Childe Rowland" is the son of King Arthur, who rescues "Burd Ellen" from the "Dark Tower." This was in fact the central chamber of a long barrow, and Arthur himself looms out of the megalithic mystery almost as toweringly as Merlin, who introduced Roland into the "Dark Tower." King Arthur and his knights go down to drink from Camelot like the King Stone and his Rollright Circle, and, as Arthur will one day wake at Avalon, so, they say in Long Compton and Little Rollright, will the King Stone turn once more into a man and conquer England. The dolmen a quarter of a mile away from the circle is called the "Whispering Knights," and is an oracle that recalls the *pierres qui chantent* of Brittany.

The kingly tradition is indeed very strong at Rollright. It is told that the witch (no doubt the ancient megalithic Earth-Goddess) of the place stopped an ancient king with these words, "If Long Compton thou canst see, King of England thou shalt be." The king replied, "Stick, stock, stone, King of England I shall be known." He took seven strides forward and came to the long barrow on the other side of the old quarry. Then the witch sang:

> As Long Compton thou canst not see,
> King of England thou shalt not be.
> Rise up, stick, and stand still, stone,
> For King of England thou shalt be none.
> Thou and thy men hoar stones shall be,
> And I myself an eldern tree.

So the king and his men were turned to stones, as we see to this day, and the witch became one of the ancient elders so abundant along the old trackway.

The chronicler Nennius relates of a stone set up by Arthur, with the foot of his dog imprinted on it, that "men come and carry off the stone in their hands for the space of a day and a night and next day it is found again upon the cairn." So the Cotswold villagers tell that the capstone of the "Whispering Knights" was taken away to make a bridge in Little Rollright. It took a score of horses to drag it down, and on three successive nights it turned over on its back. So it was replaced, and it took only a single horse to draw it up the steep hillside. Thus when Roland, the son of Arthur, made this sacred place his kingdom, he replaced an earlier deity, once the King of Cotswold. But this old king had his revenge in safeguarding his worship under the name of the most Christian knight who supplanted him.

No megalithic site is so rich in tenacious tradition as Rollright, and most of the local legends cling to the humanity of the stones. On Midsummer Eve, the day of the summer solstice and the ceremony of sunrise at Stonehenge, the villagers cluster round the King Stone and cut the eldern-tree in blossom. It, or rather she, then bleeds, and the King is seen to move his head. From the blossom is made a tea, and when you drink it you see the "Elder Mother" seated among her sweet-scented flowers and foliage, like the "Pulque Goddess" of the Aztecs, another form of the Earth-Goddess, as was Dame Elder in Saxony and Denmark and Ceres in her oak-form. As her worship decayed, she became the "witch" or the "old wife," and her powers were the same in all the megalithic centres. On certain saints' days the stones become men, join hands, and dance round in a ring. So do the fairies round the King Stone to this day and so do the stones of Stonehenge, *Chorea Gigantum,* which have the same diameter (100 yards) as the Rollright Stones, and like them are said to be countless.

In many a prowl round Rollright I have discovered that the villagers in the neighbourhood, especially the old men, still half-believe in the legends that Evans collected. A woodcutter told me that he had counted the stones three times without being able to obtain the same number each time, while he firmly believed that they went down the hill to drink at midnight. A delightful little manor-house half-way down the ridge remains forsaken because the "Stwuns" are reputed to pass it on their way for a drink. Stukeley reports that cakes and ale were the formal meal in the festal dances round the King Stone, and my woodcutter knew all about the wheaten cakes placed on each one of the stones to count them, but always one of the cakes was missing. Originally, no doubt, these cakes were propitiatory offerings to the witch, as Beltane barley-meal cakes are still offered in the Highlands to exorcise the banshee. But in the beginning they were the food of the dead, to keep them alive in their stone habitacles. From these folk-tales and village myths, flocking about the grey men of Rollright, it is possible to disentangle three strands – funerary games and rites in honour of the dead, divine kingship, and the immortal human livinghood of the stones. The portrait statues of Egypt, if they were not the originating inspiration of the whole megalithic belief in the livingness of the tombstone, yet explain this most curious belief itself, while its transmission and rationalisation account for all the variations in folk-lore.

Wold Without End (1932:38-43)

Wiltshire Downland

THE Wiltshire plateau covers so wide an area and most of the villagers are so warmly boxed away along the river valleys, that you have the illusion of seeing into distances without ever looking below you. It is as though you were striding over the clouds without getting a glimpse of the earth beneath, and yet the eye can follow the rise and fall of the fleecy masses to the limit of its capacity.This is just the effect of the Wiltshire Highlands and it enhances the liberating sense of inhabiting another cosmos from the plains below, remote from the follies and tribulations of living man. Sidbury Camp, beloved of Colt Hoare, and standing upon a steep promontory overlooking the forested country of Collingbourne across the Hampshire border, supplies an illustration. From the double line of ramparts he tells us how he could see from the Isle of Wight to Fonthill Abbey on Cranborne Chase. These ramparts crown the hill high above Everley whose 1813 church in the full Gothic style of that charming if slightly crazy period, pinnacles and battlements and all, is second-best only to the similar church at Mildenhall, the Roman Cunetio to the east of Marlborough. But what the eye sees immediately below is something very much older than even genuine Gothic. It sees a semicircle of tumuli on the bare surface of Snail Down, one more of the great archaic churchyards of Wiltshire, and, in their formality and yet correspondence with the wildness of the scene, a kind of landscape-gardening of the downs. The uneven but smoothed outline of the Marlborough Downs stands between earth and sky to the north. To the west and south-west a curve of downland, walled like the sky by the horizon, is a playground for light and shade to meet in endless combination and for the subtle, changing colours of the plateau itself. Even in August or persistent drought, the surface of the downs is never bleached and sere and wearied out like the pastures of the plains below, while on days of brassy heat the rarified air is still sweet to the lungs. On the east, a low ridge of beeches curves up to the colossal ramparts of the abandoned citadel. But to the west the bare, lone downs in tireless permutation of shape and contour, stretch on and on into the dormitory of the sun.

Bowls Long Barrow, almost in the midst of the north-western quarter of the Plain, presents a similar and even wilder scene. I made a special pilgrimage to see this long barrow, because it contained one of the famous bluestones from Pembrokeshire used in the building of Stonehenge. There could be no more telling example of the intimacy and continuity between the Neolithic and Early Bronze Age (Beaker) cultures. The remoteness of Bowls Barrow exceeds that of any other barrow in Wiltshire. Between the Wylye Valley to the south and the northern edge of the Plain and between Tilshead and Warminster, a distance of nearly ten miles from east to west, there is only one village and that is the hamlet of Imber. Yet of old, Tilshead and its surroundings were full of life. They were a lair of many long barrows and the last refuge of the great bustard that once in great numbers scudded over the Plain. Though there are still trackways innumerable, only one hard road to Imber runs through this territory and stops there. I found the barrow mutilated by a hideous iron contraption, though I was a little compensated by picking up a tiny phallic charm within its disturbed eastern end. But it would have been worth the tramp, if there had been no barrow at all at the end of it. There I was as solitary as one cloudlet in a seamless blue sky and that is what I love. How magnificently the barrow is placed! It is the meeting-place of four green roads and, though set upon only the gentlest eminence, commands the whole chalk-lands of north-western Wiltshire. The Plain spread out before me like a Pacific, wind-roughened and yet tranquil, green in the dips, gold on the crests of its rollers, while the scanty beech-groves stood out as little atolls in the dreaming waste. The tall field campanula at my feet had its purple cups filled with sun liquor, and I felt myself nearer to the primal source of energy than ever before on the downs.

The distinctions between earth and sky were half-effaced, and that enlarged being the old poets write of (certainly not the new) was within my reach. I might have sloughed off this heavy humanity and become a linnet. But I turned sadly and walked away.

One last such scene. I set off from the middle of the Plain, a little south of The Bustard Inn, to cross it as far as Urchfont on the margin of Pewsey Vale, some nine miles. This central area is as tenantless as are the high prairies to the west of it. Hippisley Cox calls the green road I travelled the authentic southern branch of the Great Ridgeway past Old Sarum, but it may have dropped south from Marden in the Vale, a little to the east of Urchfont. There used to be a huge earthen circle there which is now a piggery. I had not gone far into the wild before I saw far ahead of me the familiar couchant form of a long barrow, and, like Bowls Barrow, it is a point of rest for the sight over leagues of down. It is Ell Barrow, the lord of this region. All ahead of me was bare as far as the rich folds of land on Urchfont Hill. The head-dresses of beeches on the summits were even more widely apart than in the neighbourhood of Bowls Barrow, and you could only tell where the Plain ended by a thickening of the trees into a long smudge on the sky-line. I was like one adrift in an open boat with nothing ahead or behind me, on my right hand or my left, except the curve of the horizon, here light and opalescent, there blurred with the trees of the downland rim. Nevertheless, I know no country like this vacant downland to bring such exhilaration to the senses and yet be their peacemaker. In the composure and austerity of the chalk upland, the very thought of more spectacular country is a weariness. My disburdened mind floated to and fro in idle thought, as vagrom as the Long Ditch which meandered over the Plain to my right in the way these old boundary lines or embanked cattle-lanes of a vanished people do – Old Ditch, for example, between Knook and Tilshead – in so many parts of highland Wiltshire. The military had been shelling the Barrow and beside a piece of shrapnel I found a fossilised red-deer antler in its ditch, the miner's pick that had helped to dig it and pile the mound. Range after range diversified the level plateau, each thrusting out a tongue of land towards or away from the green road that the position of Ell Barrow beside it surely made as old as the tomb. I passed Wilsford Down and, having the two simple earthworks of Casterley Camp and Broadbury Banks on my right, sighted the Marlborough Downs ahead. Their blue serrations for the first time had limited my vision on the journey.

English Downland (1936:32-4)

Cranborne Chase

THE great way to see Cranborne Chase and to be imaginatively fired by the palimpsest of its cultural continuity is to take the high road from Salisbury to Blandford Forum. A great many people do, but they get from the one place to the other in less than an hour. You want at least a month for the journey, so constantly will you be diverted off to the right hand and left as you go. Would that I could linger over this enchanted ground for a hundred pages! Meet first the area of bare down, once congested with humanity, on the left of the road and immediately south of the great yew forest below Odstock. Six long-lived cultures have fixed their signatures to it. The first, that of the long barrows which absorbed that of the beaker and round barrow men in spite of the latter's industrial progress and taught them the holy significance of stones, the token of their goddess of earth, is richly represented. There are four long barrows here, the Giant's Grave on Braemore Down, the curious Duck's Nest in the middle of nowhere north of Rockbourne with its priest-like yew rising from the undergrowth that covers the mound, and Knap Barrow and Grans Barrow on the edge of Toyd Down, a mile above the Allen River. A complete circle of hills surrounds them from Pentridge to Whitsbury, while their track runs on past the Duck's Nest to Rockbourne Down, where on a clear day you see from Inkpen Beacon to the gap of Corfe Castle. The country is very wild about them, patched with scrub, and the three long barrows in sight of each other have a certain oddity in structure which communicates a strangeness to the setting. It seems to be the land of the *living* dead.

How many long barrows are there in Cranborne Chase – twenty, thirty? They top the crests all along our road, staring into space, the sphinxes of old England – Wor Barrow near Sixpenny Handley, excavated by Pitt-Rivers, the noble pair on Gussage Cow Down, by the Bokerly Dyke, on Thickthorn and, the grandest of them all, the Titan of Pimperne, shaped to a downlike grace. Round barrows are scattered like stars in the sky and of their age are the squareish pastoral enclosures of Martin Down and the "Soldier's Ring" in a slight hollow below the edge from which I looked down on the wastes of the New Forest, darker than their wont with rain. The next age follows a thousand years after, but here only a mile away, and witnesses the huge fortifications of Castle Ditches above Whitsbury, clouded with trees. Ackling Dyke speaks the Roman name, the village on Damerham Knoll its subject-people, Bokerly Dyke, an

England stripped of its foreign guard and desperately waiting the Saxon terror, while the maze on Mizmaze Down by Braemore carves out in turf the sinuosities of evil a thirteenth-century pilgrim must find his way through to the central hump of Paradise.

The next stop is at the Bokerly Dyke and the main road slices through it. Pitt-Rivers proved that the great ramp was raised and the steep ditch was dug after Rome had shrunk and ebbed from our downs. The defences face north to save all this desirable land from the advance of the West Saxons stationed by the Avon, and it stopped the march of ruin for thirty years. I have walked its ridge in rain and thunder and watched it winding up the naked hill in three great curves and folds as though it were the carcase of some mighty worm of legend, until it lost itself in the woodland shades of Blagdon Hill. Man has added it to the landscape. Nature has accepted it and time brought it into peace. Its serpentine course for four miles is because the fearful Britons built the Dyke upon the rambling line of a still older one. Perhaps, it is of much the same period as the boundary mark of Grim's Ditch, a sunken way between two ramps which, at right angles to the Bokerly, wanders about for fourteen miles over the downs of the eastern Chase just as I like to do, though not quite so far. Here, then, the Romanised Britons reverted to the pre-Roman manner when their masters' backs were turned. Pitt-Rivers and others have made a great to-do about the Roman Conquest, but, granted the Imperial Peace, I agree with Allcroft that the "legionaries," a polyglot assembly from all over Europe and Asia, were less highly civilised than their conquered. Even the Roman towns were nuclei of native rather than Roman life, honoured with Roman names but racked by official tax-gatherers. It is fitter to speak of the Britonising of the Romans rather than the Romanisation of the Britons, and, when the Saxons burst in, those who had lost their initiative and freedom under one military domination became the hidelings and fugitives of the next. Their decadence was inevitably swift and England was barbarian up to the landing of the Christian missionaries. Only the village community went on and on.

Oakley Down succeeds and here is spectacle indeed. Beside the road lies the necropolis of an older and longer civilisation than the Celtic or Romano-Celtic. The rough down, overhung by the shoulder of Pentride, is constellated with barrows – Wor Barrow, huge bell-barrows, the more ordinary bowl barrow and some remarkably fine specimens of disc barrows. Through the formal bank of one of them slices the Roman road on its inflexible way from Old Sarum to Badbury Rings and Dorchester. The causeway or "agger" with its side ditches remains very distinct along this portion of the Ackling Dyke, a telling illustration of the Roman bondage to rule where the hard chalk made it unnecessary. The Romans were brilliant but pedantic engineers. As oblivious of sacred tradition, the Romano-Celtic fields usurp the place of sepulchre and the contemporary road cuts through the ditch of one of the bowl barrows. Nature but not the human ages have regarded its sanctity. But still the wayfarer can feel the long shadows of Stonehenge and Avebury stretched over Oakley Down and marvel at the strength and piety of their singular religion, absorbed in death and at the same time devoted to the principle of life. They thought of Oakley Down, not as a graveyard but as a sanctuary of living men on the other side of death.

Gussage Cow Down lies over the next ridge and still the land belongs to pre-history, for the villages of the Gussages and Tarrants hereabouts are all poked away in narrow defiles and unsuspected pockets of the downs. Witchampton is the best of them in this area, being generously endowed with such antiquities as an abbey barn, a mill and fine old stone bridge, a lych-gate and one of those manor-houses in which Dorset is exceptionally rich. The valley villages of the Dorset downland show the familiar combinations of flint, chalk and brick with clusters of thatched roofs, but the architectural workmanship is inferior to that of Wiltshire, where brick, flint, chalk, pudding stone and sarsens are built in together with bold experiment and resource. Where Dorset excels is in the sounds, honeyed or sonorous, of her village names. Nevertheless, local character is stamped upon the regional divisions of the Dorset chalk. The chalk midrib between Hambledon and Beaminster gathers villages built of chalk, flint and brick. Straggling Ibberton, seen lying among its meadows from the church perched above it; Okeford Fitzpaine, with its cottages run together, and Sydling St. Nicholas by Sydling Water, with long winding street lined with whited brick, grey stone and dressed flint, tiled or thatched, are the best of them. The southern range, from the neighbourhood of the limestone, shelters mostly villages of grey stone, while Georgian Beaminster is influenced by the yellow stone of the district.

But on Gussage Cow Down is the tracing not of a village but of a town like Woodcuts. All the ages of man in society, but beyond the reach of history are written on the scarred face of the Down – a Roman halting station, the Celtic town, four round barrows and a disc barrow, a pair of uprearing long barrows that take the whole ridge into their keeping and, best of all, a great Cursus, unique in Dorset. All the dark face of Cranborne Chase, embossed with its hillbrows and of light relieving greens where the bare chalk is supreme, lies open from it to the west and north, while shaggy Pentridge, the Esau of these hills, looks closely down upon it. It is towards Pentridge that the Cursus flows for nearly four miles, two parallel banks and ditches about a hundred yards apart, and the triple and quadruple embankments of the forsaken city cut clean through it, thus revealing the former's earlier tenure of the hill. To what green altar, O mysterious priest . . . what sacred games and high processions once trod this lonely turf when Stonehenge was the Mecca of the downs? The city came with its cattle-pens and tracks and winding banks and common fields of arable, the pitting and humping of the ground where once were human houses. As the Cursus was in their days, so is the highland town in ours, a meaningless scribble on the turf.

If you continue along the main road to Blandford and take the next turning to the left, you meet the Wimborne-Cranborne road at the Horton Inn. Its landlord, standing in front of a ribald and Reformation print of monkish scandals on his wall, once told me that his ancestors were sun-worshippers. What he was referring to were the Knowlton Circles, a little nearer Cranborne, on the nave-wall of whose originally Norman Church St. Christopher carries the child over the stream, in spite of a fish to be seen taking a piece out of his leg. Only one of the four circular earthworks has escaped the plough. In the centre of its perfect round of high ramp and inner ditch stands a deserted little flint church of the early fourteenth century, standing like the moon when, foreboding rain, it is encircled by a wide aura. It is a beautiful example of the continuity of sacred sites, for, with the exception of Oakley Down, Knowlton, with its round barrows, is the only other *locus consecratus* visible on Cranborne Chase. In structure and plan it is a little Avebury without the stones, and, likely enough, timber uprights were set up along the lip of the ditch as substitutes for monoliths of stone. It also reminds you of a magnified disc barrow without the central hump. But who knows what mound or burial chamber was supplanted by the forlorn little church, whose ragged and dismantled walls now admit not worshippers but only the wind and the rain.

What faith, what passion and organised urgency flung these laborious monuments over high places hundreds of miles apart, from Salisbury Plain to the East Riding, and from Crete to Malta, Spain and Britain! As Colt Hoare said of Stonehenge – *Tantum religio potuit.* Perhaps that religious sentiment was so strong in this isolated district that the Christian faith itself bowed to it and the church was built within the heathen circle. Now all is dead: even the ivy tree is dead, whose withered arms embrace the walls of the nave and the church-tower to its summit. Over this abandoned, local site of two religions which have moved the world hangs a melancholy pervasive as its winter mists. "It hath a look that makes me old, And spectres time again."

English Downland (1936:43-7)

Hambledon Hill

THE next area of Dorset chalk is a belt of rich country (and so partly overlaid with other geological strata) running from Shaftesbury in the north to Wimborne in the south, cloven and watered by the rushy Stour. A long chain of Early Iron Age citadels, or, more properly speaking, fortified hill-towns – Hambledon and Hod, Buzbury and Badbury Rings above the eastern bank and Spettisbury Rings aloft the western bank – magnifies the natural river barrier to access into the kernel of the west country. The chalk itself admits no check nor hiatus to its majestic western expansion, for the ridge travelling from Shaftesbury and Melbury Hill is continuous with Cranborne Chase, while the watcher from Badbury's head-dress of pines, at which half Dorset looks, turns one way to the woodlands and bare chalk crowns of the Chase, and the other way to the cretaceous highlands of mid-Dorset. Spettisbury is only a fragment, but Buzbury Rings was settled by peoples both earlier and later than its builders, and the Romano-Celtic village there has partly obliterated the older ramparts. At Badbury, Arthur's men briefly stayed the Saxon advance in A.D. 520 by the battle of Mons Badonicus, and within the mighty battlements of Hod is the smaller square camp of Lydsbury, called Roman by many. It is much more likely to be an imitation Roman fortification built by the despairing Britons whom the Romans left to be devoured piecemeal by the Saxon dragon.

Of the hill-towns of the Stour, Badbury is the most comely, the most handsomely designed to the contours of its yet shapelier and isolated hill, while Hambledon excels in sheer magnificence. Though the arterial road, the Ackling Dyke from Old Sarum to Dorchester, passes Badbury, there is no sign of Roman work there except that rare thing, a group of Roman tumuli, recognised by their conical form. Bronze Age bowl barrows and a disc barrow share the ground with them, and Nature has pressed all these ages of man, those of peace and of war, into one harmonious masterpiece where contemporary man, slipping for a moment out of the clutch of his own age, more menacing than any in the past, may drink deep of a unity and repose transcending time itself.

In all my survey of the chalk highlands, I never climbed a hill of more grandeur than Hambledon, though Eggardon, the door to Devon, and Bow and Butser, the double gates to the South Downs, are its equals. To begin with, Hambledon affords a perfect illustration in shorthand of the natural tendency of the chalk to form wedges out of a central core, tapering to the extremities. Three flying buttresses or spurs converge to a plateau little over six hundred feet above sea-level. Yet the dizzying steepness of the sides to the west, north and east, and the beautiful lines of these outward-bound tongues of land to the north, to the south-east in the direction of Shroton and southward towards Steepleton Iwerne and the Hanford Gap, give a mountainy impression of Hambledon's crookt, oblong, slighty up-tilted plane. The power and proportions of Hambledon's sculpture are not blurred by a burst of wildness in its natural vegetation. The triple-scarped drop on the east is powdered with tenebrous yew, gnarled thorn, the gleam of whitebeam and tracery of ash; primitive scrub darkens the folds of the south-eastern wing and, as Mr. Heywood Sumner noted years ago, still the traveller's joy flings its downy awns over the wind-riven bushes.

Lastly, the human inscriptions from a remote past are bewilderingly rich and varied. The natural ascent is by the Shroton spur where you look down upon the flanks of the spur, furred with trees, that makes for the Hanford ford of the Stour. Rise higher to the more level ground where the two spurs unite, and Hod Hill with its bald crown garlanded with green rings lies below, dwarfed by the bulk of

Hambledon. Here is where the faint, low-banked Neolithic camp is pencilled, with a massive long barrow for its contemporary high in the middle of the later camp on the saddle of the hill. Ponderous scarp-to-scarp defences rear their walls across the heads of both these spurs and, past them, the great camp with its battlements in triple tiers, flows right round the edge of the hill, with two cross ramparts dividing it into three sections. They clasp the hill-front in curving, gliding folds like the coils of some fabulous serpent and block the margins of the sky, so that even the round barrow along the northern spur is belittled. They are the work, constantly brought up to date, of successive periods from 500 B.C. onwards, the huge outlying ramps of the southern end being the latest. When the work was done, unaided by the natural advantages of the hillside, the toilers looked across the valley to Shillingstone Hill with its dorsal crest of trees, and the vertebral summits of the hills away towards Devon; to the mists of Mendip far away in the north-west; to dark Shaston and pale green Melbury and down to little Hod behind them. They knew themselves to be impregnable. Yet the camp was deserted, unavailing against either Roman or Saxon, and leaving to unborn generations nothing but delight in its union with the still grander work of Nature and an innocent awe at the shadow of its greatness. This was an achievement better than if Hambledon had beaten back all the power of Rome and of the barbarians that flooded England after her fall.

English Downland (1936:47-9)

The Isle of Axeholme

IN the course of my desultory but extensive reading, I hit upon scattered references to a mysterious region I had never previously heard of – the Isle of Axeholme, in Lincolnshire. I learned to my astonishment that this "Island" had escaped the universal village tragedy of the Enclosures and was still functioning as a small galaxy of village communities practising the open-field system. The main attempt to enclose the Island occurred, so far as I could trace the scanty records, in the seventeenth century, when Charles I, the Duke of Bedford and the Dutch engineer of the "Bedford level," Vermuyden, put their heads together to make exorbitant profits out of the scheme for the drainage of the fens. Partial reclamation of the Axeholme fenlands had been undertaken in the Middle Ages, and the land rendered very fertile by the "warping" of the Trent and deposition of its silt upon the fields. Each of the partners to the scheme was to have a one-third share of the reclaimed land. Rioting broke out all over the fens, but the natives of Axeholme were alone successful in retaining their ancient rights. They had resisted the Enclosure Commissioners, "dispeoplers of towns, ruiners of common-wealths, occasioners of beggary, cruell Inclosiers whose Adamantine hearts no whit regarded the cries of so many distressed ones."

Arthur Young, the ablest of the advocates for enclosure, witnessed and partly confessed to the bitter fruits of his triumphant campaign. This is how he wrote of the smallholders of Axeholme: "They are very happy respecting their mode of existence. Contrivance, mutual assistance by barter and hire enable them to manage these little farms. A man will keep a pair of horses that has but three or four acres by means of vast commons and working for hire." Every cottager possessed a common right, and no rights were attached to the land apart from the cottagers. Rider Haggard, whose passion for the restoration of English agriculture was the guiding principle of a disinterested life, had not passed over this well-filled pocket of antiquity. It is, he wrote, "one of the few places I have visited in England which is truly prosperous in an agricultural sense." The rest of the facts I unearthed concerned the high average yield of wheat and oats and similar agricultural information. Most of the cottagers kept each a cow, a horse and a few pigs; the strips were ploughed by co-aration; holdings varied from half an acre to two hundred acres, and it was possible for the smallest cottager to acquire by his diligence larger holdings up to that amount. By such means they were enabled to face the long winters without hunger or pauperism and to stand the strain of agricultural depressions "exceptionally well." They had preserved the spirit of the old system, its social equality, its principles of mutual aid and its aim at the maximum gross produce of food rather than its maximum net profit.

But all these records, being of pre-War date, might well have been chronologically unsound. I could discover none relevant to our own period, and there was no way but that I should travel up and see these remarkable islanders for myself. Only by personal enquiry could I find out whether the era of big mechanized farming had overwhelmed them, or whether this little band of economic dissenters and rural conservatives had ridden the flood under which so large an area of agricultural England lay drowned. It seemed to me that these islanders were more exciting to visit than the Anthropophagi and men whose heads do grow beneath their shoulders.

I reached my goal in the extreme north-western corner of Lincolnshire. I had hardly expected to discover an island in the middle of the Midlands, and yet the extreme insularity of this little patch of the living past was the first impression I received. What the natives of Axeholme call themselves is the "Isleonians." The Isleonians. Not some mythical tribe of wonder-folk in the archipelago of Tir-na-Nog, nor a snapped-off twig of the Atlanteans, nor a company of demigods that had made some haven of the chartless Hesperides, but a handful of husbandmen living in a few villages belted by manufacturing districts. Scunthorpe to the east, Doncaster to the west were near enough to give the term "island" an intensity of meaning that might have pushed Axeholme into the middle of the Pacific. My eye swept the horizon from the low ridge between Haxey and Epworth, the capital, and saw the smoke from the factories smutching the sky. So, a hundred years ago, the new manufacturing towns were creeping over the northern countryside like a fungus over the leaves of a plant. I had to pierce part of this industrial fringe to fetch up in Axeholme, so that "island" is the word. An island its rivers declare it to be – Trent, Idle, Torn and Don – and they have been the boundaries from the remote past. It is curious that Axeholme and Lincoln are the only districts in Lincolnshire which retain traces of Celtic place-names.

Had I been mapless and blind to finger-posts, a sign, a strange writing on the land would have told me at once that I had reached the country of the Isleonians. Its surface was not patterned into the quilt of our familiar vision: its hide was no longer the panther's but the tiger's. The land was striped, and these stripes, making a Jacob's coat of many colours and laid across the soil in narrow bands of more or less than half an acre in extent, are locally called "selions." I saw no baulks between them, but each ribbon of "champion," to use old Tusser's term, possessed its own name. Blackamoor was one acre and twenty-seven perches, Scab Flat thirty perches, Short Crabtree thirty-two, Godspeed thirty-six, and others of slightly varying sizes possessed each its Christian name – Foxholes, Meadow Piece, Near Shore Field, Far Shore Field and Tassel Croft. They seemed to personalize every blade of corn.

This novel sight might have repelled one accustomed to the chessboard effect of the countryside, especially the small fields of the Home Counties. But to me whose chief preference are the wolds and unenclosed sheepwalks and have so been taught to value the structure and anatomy of landscape more than its clothing, these open, banded fields proved congenial. The rural scene of Axeholme has no salient appeal to the eye, and hedged fields would have made it very dull. The selions gave it its best chance, especially along the slight ridge between Haxey and Epworth, where the strips, conditioned in their shape by the bend of the road, curved up the slope in the form of an inverted S. The variety of crops grown over so restricted an area also lent the scene a wide diversity in colouring, entirely lacking, say, to the pasture of the Vale of Aylesbury.

The villages, too, of this miniature of the traditional England, whether sprawling like Haxey or compact like Epworth, the birthplace of John Wesley, were built of local materials, red brick of a warm shade with wavy roof-tiles of a lighter red, and in brilliant contrast with the black and white windmills, round-knobbed, Dutch-looking, and with a decorative ball on the tip of each slender shaft. In the old days, the villages must always have reflected the geological strata, the natural configuration and even the vegetation of their particular localities. They fulfilled the aesthetic law of fitness to environment. The village community was so wedded to the land that the varieties in the structure and productions of each village faithfully transcribed the regional diversities of English landscape. Only the days of mass production have given scope to the red-brick villa equally to scale the Pennines as to edge the Saxon shore.

I had never in my life seen so many different crops grown over a small area. Flax, remarked upon by Arthur Young, and hemp are no longer cultivated. But there are oats, wheat, rye, barley, cabbages, potatoes, sugar-beet, red beet, celery, carrots, beans, peas, onions, parsnips, cauliflower, leeks, even red and white poppies, whose seeds were dried and sold to the chemists. All the cereals grown in England are raised by the Isleonians and most of the vegetables. One of the theoretic arguments of the land-grabbers against the old agriculture was that it was uneconomic. What I could see for myself was that it was intensive as opposed to the extensive system of the Enclosures. Not a clod of earth but was cloven. Except for a minute percentage of pasture land the whole island is arable. Its natives had applied the maxim of Cicero: "Nothing is more excellent than agriculture, nothing more productive, nothing more pleasant, nothing more worthy of free men." But the common pasture and common waste of the pre-Enclosure village had been shorn from them. Lacking the pasture to feed them, the cottagers were no longer able to keep cows. Up to the time of the War, the strip-owners used to club together, ten or more, to pay a man for herding their beasts, taking it in turns to relieve him every Sunday. The herds, up to twenty to a herdsman, used to graze the rich grass bordering the byroads, until the motor car came and with it the byelaw that cattle on the roads must keep moving. Nowadays, pigs and poultry are the only cottage livestock. Even the cottager who owns or leases no more than half an acre has his pig, and usually more than one. The Isleonians are "fond of a pig," my friend the miller told me, and there are plenty of waste potatoes for them and mills to grind for them. But if they have lost their cows, they still, so to speak, keep the milk in the family. The supply is in the hands of two or three small dairy-farmers on the borders of the island who sell all their milk in the district. That common, ironic spectacle of our countryside – the milk-lorry carrying off the rattling milk-churns to the big town – is unknown in Axeholme.

Genius of England (1937:118-24)

Man and Nature

WHAT has concerned me has been the partnership between man and the ground beneath his feet. To violate it means his excommunication from the spiritual fellowship conferred by the bond. I use the word "excommunication" in the sense of expulsion from a communion which is definitely religious in spirit. The covenant is a relationship to the land and the land a little word that translates into secular terms the earth-goddess of our ancestors. The

Puritans made their revolution by the confession of faith that each individual man stood in a personal relationship to God and was answerable to God alone without the intervention of professional intermediaries. This is a true analogy to man's relationship with the earth which, if not God, is the mantle of God. Our intercourse with the earth is multiple in form; it stretches through a long series from contemplation to the satisfaction of man's daily necessities. Rural England is or was a living witness to the fruits of that intercourse, whether manifested upon a page of the Oxford Book of English Verse or on the furrows of the ploughed field. Rural England is not nature but man's transformation of nature into something that expresses them both in unity of being. Art has been recently called "a quality of form and rhythm," and that definition so perfectly renders man's handiwork upon English earth that the countryside of England acquires by it the highest title of distinction the world can show – that of a work of art.

The pulse of wild nature beats to the breath of God, but man has refined and matured and sweetened that untutored heart and so redeems his own from the sorrow and futility writ large upon the history of his civilisations. It is not hard to decipher the wording of that portion of the covenant he must not break. It blesses every deed and thought that makes yet more fair the inheritance a thousand generations have bequeathed to us. But it curses with blindness of understanding, with exile from communion and discord of mind any man or society of men that distorts the form or snaps the rhythm. There is no curse more bitter because our universal mother is this very earth and to do her dishonour, to be lacking in piety towards her, is to be rudely cast out from the first steps of the great stairway leading to the courts of heaven. Well did Blake say that to put a bird in a cage puts all heaven in a rage. In the varied business of each and all of us with soil and beast and bird and tree and the nature that is also within us, it were well to avoid the wrath of heaven that taketh away the peace the English land bestows upon us, but that the pride of empires and machines and our industrialised society have never given and will never give.

Through the Wilderness (1935:282-4)

Craftsmanship and Nature

THE mysterious relation of craftsmanship to Nature has a supreme example in the achievement of regional architecture. Neither its prodigal variety in forms and materials nor its developments through the centuries obscure its primary fidelity to its particular rock. The reason for the diversity of the English villages is the varieties of our native rock, and the English village is a better handbook of geology than any printed text. It is not only a guide to the strata underlying it but an exposition of their qualities, texture, capacities and natural vegetation. An observant man does not even have to look out of the window of a cottage interior to tell what kinds of soils surround it, what kinds of crops it bears and even what are its natural features. He can guess the landscape by looking at the roof. I can test the geographical and vegetational changes taking place between the oolite and the lias down from the north-western Edge of the Cotswolds to the Valley of the Warwickshire Avon just as well by going from village to village as by studying the landscape and tapping about with a geological hammer. The transitional architecture from village to village exquisitely registers and translates into vernacular terms the hidden history of the earth on which the homesteads rest.

In the same way, the rather severe rectangular lines and sharp angles of the Cotswold house render into the human medium the linear composition of the wolds, while the discontinuous and fantastic outlines of the liassic hills are expressed in the broken surfaces, the medley of materials, the happy craziness of design in the brick, timber, plaster, thatch, tiling and sandstone of the cottages in the liassic Vale. And because there are no villages in all the world to equal those of England between the 14th and the 19th centuries, so there was no breed of craftsmen like the English who built them. Variety is the English genius of craftsmanship; variety the English genius of landscape. I can at this moment, sitting in my chair, picture to myself four totally different types of cottage limestone-slatted roofs alone, each true to its geological environment, each within fifty miles or less of one another.

If, again, after the geology lesson of walking up the village street, the craftsmanly minutiae be attended to – the chamfering of a waggon in the shed, the carving of a corner-post or of a piscina or a poppy-head bench-end in "the decent church that tops the neighbouring hill," the ogee curves to the "guide" of a shepherd's crook, the wrought-iron work of a weather-vane or a chest either in the church or a farmhouse, the moulding of a dripstone over a cottage window, the geometrical pattern of the thatch below the ridge-board or (if the roof be stone) the "valleying" of the angles of intersection, the setting of the lights to the wall-space and of the houses to the lie of the land, the siting of the church in relation to the secular buildings, the mingling of colours in a cottage garden, the cornering of a thatched wheat-rick, some blue ware on a dresser, the shape and colour of a mug at the pub, a carved settle opposite it, the ball-flower ornament over the chancel arch, the proportions of a pigsty, the raftering of a barn, the brasses on a horse's martingale, a corbel-table along the nave, the slight batter of a dry-stone wall, the harr of a field-gate, the pewter inlay of a butterfly bobbin on a lace-pillow, the finial at the gable-end of a grange, a carved boss in the groining of the church-porch – if the multiplicity of these details be taken in and a Nelson eye be turned to modern intrusions; if, again, another village ten, twenty, fifty or only a couple of miles off be remembered as totally different in its materials, its forms, its mannerisms, its styles, even of the tools (which also will have different names), then it will appear that, if we were a nation of shopkeepers in 1800 and are a nation of card-indexers and committees to-day, we were a nation of artists in 1500, 1600 and 1700.

The English Countryman (1942:56-8)

The Craftsman

AND that question I had asked myself during the Plumage Bill agitation – was the acquisitive sense in man a natural inheritance or the product of historical and economic factors favourable to its development at particular epochs in time, to it craftsmanship was the answer. The craftsman's relation to nature was non-predatory from first to last, from raw material to finished product. He did not conquer nature but married her in husbandry. Or, if I cast my mind back to the old controversies that had raged during the Battle of

Diffusion, here too I was fed and satisfied. By means of craftsmanship, art was "in widest commonalty spread", both on the wolds of Gloucestershire and in the caves of the Dordogne. The only difference was that the primitive had become civilised without turning his back upon his birth. In craftsmanship, there was no division between capital and labour, master and apprentice. The craftsman had no need to turn aggressive, to fly to arms. He had inherited the natural peaceableness of the primitive.

What had been put in the place of craftsmanship? I looked about me and what I saw was ruin, ruin everywhere, in fields and buildings and workshops. The great city had replaced the cathedral or the market town; but was Manchester a better place than Chipping Campden? Was the democracy of the village in the Open Field System, which had both fostered and been served by the craftsman, inferior in final value to the democracy of the vote? A chain of little local aristocrats, of master-men and apprentices potentially so, seemed a definite answer to a proletariat.

True, the craftsman was not progressive; generation by generation he created anew along the great line of tradition. Yet I asked myself whether this way was not nearer to eternity and to nature both than is the way of progressing into the void. The whole idea of progress as we interpret it now, is a kind of slavery to the strip-film notion of Time. Is the notion of great craftsmanship less real to me because it is no longer present? Not so. The self is timeless. So, leaving the Cotswolds and returning again, I went my way, walking a lonely road, but one that I felt to be solid ground, well-paved and signposted by the historical method.

Remembrance (1942:83-4)

Goodchild of Naphill

YET among these villas exists what is perhaps the greatest surprise in the Chilterns, and so a contrast indeed, a pearl, a wonder. No less a personage lives by Naphill Common than the very last of the Chiltern handicraft chair-makers – and him no book ever mentions. Edward Harold Goodchild has his proper flint cottage and next door to it his long low weatherboarded workshop with its sagging timbers propped up on poles. This lengthy interior is quite literally stuffed to the ceiling-beams with chair-parts – arms, legs, seats, splats, back-bows, skeletons – a glorious gallimaufry of everything to do with chairs, so that he and his assistant have barely room to work and move about. How the hens find a way in is a mystery. An armoury of tools – spoke-shaves, scrapers-in-stock, travishers, adzes, cleaning-up irons, mortising gouges and their fellows – hang from the walls or litter the benches. Some have been designed by himself to pick up the grain of the wood. Beside them hang designs in paper and highly decorative banister-splats. I love a workshop like this as well as any landscape: indeed, I find an affinity between the one and the other. The smell of the shavings, the feel of the timber in the making, the vast serviceable untidy litter exhilarate me like wine. A workshop like this is a fundamental piece of reality, the dynamic link between Nature and Man, and the spectacle of what Man can do when he is in tune and in touch and in play with his mother-earth is an indescribable refreshment and release from the hideous unreality of modern life. Though the crafts are killed off one after another, there is a sense of the undying about them, of a sweet permanence that has solved the problem of adjusting means to ends, life to environment, beauty to need. The gigantic muddle of modernity, the consequence of losing ends in means, is shamed before this spectacle of creative order emergent from chaos. You know that this is what really counts in life – this honest man making handsome, solid, craftsmanly chairs out of the trees in his neighbourhood, oak, walnut, yew, beech, apple, pear, elm and chestnut.

His tools are, of course, more elaborate than those of the bodger, the number of the turning chisels is greater and, like Samuel Rockall (who once made me a Windsor chair), he uses the fly-wheel lathe. He steams his own backs from a tank in his orchard. As among the chair-leg bodgers, every process from A to Z passes through his own hands, and this is one reason why the rhythms of craftsmanship bear so close a resemblance to those of Nature and none to the routines of mechanical production. Mr. Goodchild has been a chair-maker for twenty-seven years and of course his father was in it before him, though he only made seats. The last time I was in his workshop I ordered an arm-chair of yew-wood, chairs in this difficult wood being some of his finest work, partly from the dark glow it acquires through persistent rubbing by home-made beeswax and linseed oil or, surprisingly, by a tallow candle. My chair with its cabriole feet from a Queen Anne model and elaborate banister-splat magnificently shows the glossy piebald wooding of the yew. But making chairs from yew-wood is no joke, the worker frequently suffering from migraine or dizziness on account of the poisonous exhalations.

Mr. Goodchild takes his designs, modified for the occasions, mostly from Sheraton, Hepplewhite, Windsor and Ladder-back models between 1700 and 1780, and on the occasion of one visit to him I made a truly romantic discovery. A dozen or so years ago I bought at Heal's a pair of "Gothic pattern" arm-chairs – chairs, that is to say, whose backs resemble one type of elliptic Decorated bar-tracery in fourteenth-century windows. He was showing me some of these backs and, one thing leading to another, I found out that he had been the maker of my Heal chairs. But his spirit is remote from that of the art student who sets an easel down before an old master in the National Gallery. He approaches his masters as a master himself in the tradition of English chair-making, an initiate by right of descent in the calling of a craftsman. Nothing could better illustrate the datelessness of the old crafts than this inward continuity. Their association with the past is purely arbitrary; they are back numbers only because a barbarous age has no use for them. Country crafts can never be rightly understood except as being essentially independent of time. There was no reason in Nature, a natural love of beauty and human need why they should not have gone on for the duration of man's tenure of the earth, since the generations of craftsmen represent a chain of re-creation on the exact analogy of the poet's, "Thou wast not born for death, immortal bird." Their virtual extermination in our own time has been sheer murder by an evil economics; it has nothing whatever to do with replacement or internal decay. Theirs has been an innocent death and the blood is on the head of the killer.

Chiltern Country (1940:86-8)

Samuel Rockall

THERE is only one specifically downland handicraft that to-day survives the wreck of craftsmanship at the Industrial Revolution and its slow foundering ever since. Hurdling is not a prerogative of the chalk, though once much practised on it, and the one turner left in Berkshire lives on Bucklebury Common, some miles off the Downs. But chair-leg "bodging" is an authentic industry of the Chilterns, and to the best of my knowledge is traditionally pursued nowhere else in all downland except among the highland beech-woods in the south-western corner of the range.

Bodging is on its last legs: none of the present masters have apprentices, so that when the present little company of bodgers, all of them old or oldish men, have passed to the country churchyard, there will be none to take their places. Machinery is, of course, one reason for the decline, the production, that is to say, of chair-legs of inferior quality and design at a superior speed in the making. A machine-made chair-leg is at once betrayed by the stiffness of its appearance, shallow mouldings and lack of invention in the turnery. The scandalous rates of pay are a yet more potent cause for the imminent extinction of a craft that, persisting in the traditional designs, turns just as good chair-legs nowadays as are coveted in seventeeth- and eighteenth-century examples of gate-leg and other tables. A bodger working ten hours a day for six days on end can only make thirty shillings a week, and it is evident that the only reasons why the bodgers that still tread the lathe on the hills do not desert to the High Wycombe factories, where they could easily secure more highly paid jobs, are partly the inborn conservatism of all rural crafts and partly their love for their free and satisfying work in its natural environment.

What wonder? The setting of these craftsmen's homes is as romantic as it well could be. One of them lives among the hanging cherry-orchards of Stoke Row. The bodgers that use the primitive pole-lathe are most of them itinerant and pitch their temporary "hovels" among the aisles and in the leafy seclusion of the beech-woods. Another, Samuel Rockall of Turville Heath, who uses the treadle-lathe, has a flint cottage that looks directly out upon a common of furze, bracken and stunted groups of trees. In the Chilterns, the effect of grandeur and spaciousness is gained only when mist clouds the network of the tree-tops, half shrouds and half reveals the valleys, folds and bottoms, and suggests the illimitable by means of form and vegetation rather than by distance. When the mist ramparts Turville Heath, it appears as wild as a parcel of prehistoric England, as withdrawn from civilisation as the stage of a fairy story. The entrance to it is of an as unquestionable a greatness as could be found anywhere – a towering avenue of ancient beech, sycamore, oak and lime with far-spreading branches and massive boles, most curiously wrought. Some are like baroque columns in clusters, others are richly fluted and bossed with huge carved nodes whose callosities resemble Gothic ornament in the grotesque. The buttresses that support these great trunks have their feet, or rather talons, deep in vivid moss.

A little nearer the Fingest Valley, winding down in a series of advancing and retreating folds, lives another bodger, Mr. Bartlett. His house lies back along a lane almost impassable in winter. At its opening an embanked and circular pool is overhung by a giant oak whose lower branches vein the untroubled water. A thousand trees are marshalled on the farther bank, and under them the wild snowdrops chastely announce the riot of spring. It is right that this seclusion and privacy should encompass the ancient craft, since only by lack of contact with the world has it endured and maintained its hold upon the tradition that keeps it ever fresh and new. These natural presences are the forms of that past which mothered all such crafts and still protects the immemorial woodmen from falling victim to the machine.

The one that I know best is Samuel Rockall, and I can see him now in memory's eye toiling up the steep on a veiled winter's day at the head of two wagons loaded with the timber he had just cut for his trade. In the misted landscape, his blue eyes, flushed cheeks and reddish hair were the only colours of the scene. An artist might have drawn him thus, but he would have left out his representative significance. The true craftsman controls and executes all the processes of his craft from the raw originals to the finished product, no matter how many they be. He is thus divided by a cleavage absolute from the one-man-one-bolt system of modern minutely subdivided industry. That is why the rural master-man remains by the law of his being close to nature. He is not merely surrounded by nature; he not only takes his tools and materials from nature, but he repeats the ordered unfoldings of nature from the seed to the flower, from the grain to the ear, from alpha to omega. This is the secret of good craftsmanship and the condition of its blossoming, that the man shall take the fruits of the earth from the hands of nature, and with his own hands transform them into the final form he destines for them, to be at once useful for the needs of his fellows and pleasurable to their eyes. From mast to tree: it is the same thing over again but on a new turn of the spiral of creation. Samuel Rockall buys and axes his own trees where they stand in the woodland, and so his craft is organic. He conducts all the operations from the tree in the forest to the chair by the fire, and like a magician wills nature to come out of the weather into the home.

The timbers are unloaded in front of his hovel, sawn by him into lengths and stacked in the smaller room made by the partition. The larger room in the hovel is so deep in shavings that the foot makes no sound, and is so crowded with the instruments for converting logs into legs that the eye is bewildered. Chopping-block and splitting-block stand like islands in the ruffled lake of shavings; shelves hold chisels of every shape and edge; draw-knives hang on the walls; beetles, axes, chair-legs and piles of logs lean against them, and the long treadle-lathe occupies the whole of one end, with a little wooden windmill by the great wheel to amuse the children while the work is proceeding. This is one of the toys that Rockall has made for his family when he is not bodging, making chairs for special orders, bottling fruit, grafting crab-apples, gardening, sharpening saws and cobbling his children's boots. He is a living embodiment of Blake's "Exuberance is all."

The first process in the workshop is to split up the logs with a woodman's axe and a very heavy beetle, the measurements being done by eye. The bodger then moves on to the second block which stands higher for the finer work of chopping, or rather stripping, the split lengths ready for the draw-shave horse. He usually cuts about forty trees in a course of fellings, and these, being "thinnings" or timber cut to prevent overcrowding, do more service in the preservation of the woodlands fallen than standing. Once, as he chopped away, talking the whole time in his brisk, cheerful, birdlike manner, he told me the story of how he had obtained the chopping axe he was using, a broad, finely curved blade with edge of tempered steel and a handle short enough to be covered in the hand. He had found it buried in the woods years ago when he was felling trees, and he conjectured, by what divining powers I know not, that it had lain there for sixty or seventy years. Nevertheless, he put an edge on it again in two hours' work, for Samuel Rockall is a marked man in all the region for his prowess in putting an edge upon saw, chisel and axe. He thought he knew how this chopping axe had got hidden in the woods. In the old days, the bodgers used, as a respite from their arduous work, to set aside one day in the week for a drinking bout, since, as the master of a body of them was wont to say, "you work like hosses and drink like hasses." It was the custom, when a new-comer failed to stand the company a round, for the men to say, "Mother Shawney'll be after you," and, sure enough, Mother Shawney would for a time remove one of the novice's tools. The axe was hidden but, like the squirrel's booty, was never recovered.

The draw-shave horse is a long trestle with two cross-pieces, the lower one the longer for the bodger to rest his feet. He seats himself upon it and fixes the rough chair-leg between the cross-pieces by a row of iron teeth fastened to each bar. He ties a "leather" to his chest and grasps the handles at right angles to the blade of the draw-knife, the right with the palm down and the left with the palm up. The action of the knife is upwards and sideways, so that the shavings hit the bodger's breast. The process by which the leg, shifted from one hollow between the teeth to another, is shaved and smoothed is incredibly swift and energetic, and, when half the leg is finished, he tosses the other half to him in a second and is whittling off the slivers as quick as the eye can follow him. The sawing horse is a smaller trestle, between the forks of which the shaven leg is sawn according to what type of chair – and there are about a dozen variations upon the old wheel-back – and to whether it is to be leg, spindle, side- or front-stretcher. For measuring, he removes one out of the array of rectangular measuring-boards hanging on the wall.

There is still one more process to come before the consummation of turning the wood. The chisels have to be "sharped," the upright stands for holding the chair-leg to be adjusted, the right "rest" for the chisel has to be selected from those hanging by the lathe and a new "centre" fixed according to the kind of leg to be turned, all before the foot is pressed upon the treadle and the fly-wheel revolves.

The depths of the mouldings in the leg are regulated not only by the pressure of the chisel bevelled on both sides but by the pace of the treadle which stretches the whole length of the lathe. The revolutions are very rapid for the planing and shallower turning and slowed up for the deeper incisions. So like magic is the turnery – a triple ball-turning both concave and convex appearing with a suddenness that startles the eye – that the watcher is bound as by a spell. Spouts and ribbons of shavings stream off the revolving chair-leg in all manner of fantastic shapes, spiralling, twisting into coils, flying overhead and falling about the bodger's shoulders, or shooting out in little straight jets and gushes, as though the wood was not only animated but shedding one skin after another in a high state of tension. The impression is conveyed that this was the way that worlds were made, thrown off each in its spinning entirety from the Great Magician's rod. The bodger begins by marking the places to be turned and, immediately that the concentric ridges and grooves have been made in one of them, the surface is planed on either side. In this latter movement, a number of swiftly revolving rings encircle the leg, melting into one another towards one end of it or the other and these vanishing, insubstantial, madly whirling circles enhance the sense of a ferment of creation. The work as a whole is so rapid that a complete chair-leg manifests itself in about two minutes, perfectly finished and ready to take the seat of the chair. The automatic lathe, of course, revolves at an even greater speed, and by thus passing out of human control must produce the effect of mechanism upon the object turned. Where the human will is predominant throughout in the bodger's hovel, the observer becomes conscious of the subtle blending between the working out of traditional forms and the workman's independent skill and personal initiative. Each is ineffective without the co-operation of the other.

Samuel Rockall was happy to explain why he preferred the treadle- to the pole-lathe. The one advantage of the latter is that it is portable and so can accompany the wood-bodgers from one temporary hovel to another. It is similar in appearance to the treadle-lathe but is without the heavy fly-wheel whose office is served by the tapering and flexible pole of ash, willow or fir, about ten feet high, which with its butt fixed in the ground or a square opening in the shed reaches slant-wise over the top of the lathe. A slender cord is attached to the thin end and this is twisted singly round the chair-leg before it is fastened to the lathe. When set in motion, this contrivance turns the leg both ways by the winding and unwinding of the cord, and it is only on the inward turn towards the bodger that he applies his chisel to the wood. The fly-wheel, on the other hand, causes one-way revolutions so that the bodger has no need to pause in his turning. All the same, though he is thirteen years older than

Samuel Rockall, who is over sixty years of age, the bodger at Stoke Row can turn a chair-leg on the pole-lathe just as expeditiously as Samuel himself. This man of metal turns the hardly credible number of three gross or four hundred and thirty-two chair legs in a week, working fourteen hours a day with intervals for meals. Lest some orthodox trade-unionist should hap to read these words and go white with indignation, it should be gently pointed out to him that the bodgers have preferred the mastery of their time, their labour and their product to the modern desiderata of fixed hours and higher wages in a factory where the master-man is degraded to the servant who commands neither his time nor his labour nor his product. It was not only exploitation which made the demand for fixed hours of and higher wages for labour irresistable; it was the distastefulness of the work itself from which had been emptied the pleasure that independent creation alone can bring.

That this is not illusory is shown by the example of Samuel Rockall himself. He works all day long for a beggarly pittance as a return for an output of such superb workmanship that it cannot be distinguished except by the cleanness of the wood from the highly priced antique. He works so hard that his only recreation is a change of work. Notwithstanding that four persons have to live for seven days upon his thirty pieces of silver, he is one of the happiest men I have ever met, not only by temperament but because he is continually occupied upon things that are useful to others and beautiful in themselves, that are made out of the nature at his door and demand every ounce of his skill and experience. He has deliberately considered that these factors are more worth while to him than those that are prized by millions of urban workmen. Once, he found that he could not possibly make both ends meet and so took on an "improver's" job at High Wycombe behind the power-lathe. But he could only stand it for five months. He pined for the open free life, his variety of occupation and what may be termed his home-work. "I felt shut in," he said, "like a bird that's put in a cage." So he voluntarily went back to his smaller earnings, thus reversing in his own person the universal tendency of labour to flow from the country to the town. The same principle was at the back of his preference for the treadle- over the pole-lathe. Because the pole imparts a suspicion of "tremble" to the chair-leg, it is unable to accomplish such fine and varied turning as the treadle-lathe, and Samuel's ambition was to turn the finest chair-legs over all the

Chilterns. So he discarded the pole for the fly-wheel, just as he abandoned the power-lathe of the town for the treadle-lathe of the deep, untroubled country.

Yet the shadow of a sigh stole into the brightness of his face when he showed me the grandfather clock that he and his brother had made with carving and ball-turning round the face after a Jacobean design. Beside the clock hung a bellows of better proportions than the one, a century old, beside the open fireplace. He had made this as well and chairs and toys and bobbins and many another handsome and serviceable object in oak, beech, walnut or laburnum wood, not to sell but to entertain his family and soften its economic asperities. His just perceptible sigh was not for them but that the odd moments of a lifetime had not been more frequent for him to indulge his creative bent outside the limits of bodging. Like all true masters he took great pride in his tools and had made several of them himself. Like many other old countrymen, he was a storehouse of reminiscence. Always busy, smiling, anecdotal, kindly, eager and absorbed, he seemed to point the moral that the only way to a happy life was to do and to be exactly the reverse of what our present culture or lack of it regards as desirable to that end.

Certainly the bodgers – and Samuel Rockall is not the less representative of them because he is *primus inter pares* – are happy men. They are open, courteous in their independence, quiet and assured in the midst of their arduous labours and the rigours of their heavy, consistent output. They seem very old-fashioned because their peace is unknown to the age they live in, their content an anachronism. Like their once fellows in a thousand other crafts, they are master-men. They know their place in life, they are easy-natured, affable men because they are aloof, high up on their hills, from the the stresses, strivings and self-defences of the valley. They work at incredible speed but they never hurry: they work much harder than any worker so much as knows the meaning of in modern industry, but they never complain of their long hours, even of the miserable return they get for them, because those hours are for ever varied, interesting and fruitful ones. They do not want to step into anyone else's shoes, so well-made and serviceable are their own. Modern industry menaces them and will devour their children, but their work is their castle and its domain the aisled woods within the rampart of the hills. Meeting all men on their own ground, they are secure even in the threat that is aimed at their very existence, and, when they die out, it will be with the satisfaction of having done well both by nature and by man.

The English Countryside (1939:84-93)

Pike of Tinhead

COUNTRY memory still enshrines the reputation of great wheelwrights. One of them was Pike of Tinhead, dead for many years. I was moving about in Wiltshire with a small party engaged on a farming survey. We came to a very lonely farm on the high Downs near Imber which, as I knew it years ago, was one of the loveliest and loneliest places in the world. And there in a cart-shed was a waggon of Pike of Tinhead. The farmer was a yeoman, and because he was a yeoman he had in his yard a wheatstack on staddles joined by heavy timbers – there were as many as six staddles on one side of the stack. At least I can imagine nobody else but a yeoman staddling his rick, and though I saw thousands of other ricks, this was the only one that was staddled.

Memory again; this yeoman did so because his forefathers had always so done. And why? Because they had discovered by experience that it was the proper way to aerate the rick and keep the rats and mice away from it. I wonder how many tons of wheat and thousands of pounds are annually lost by building ricks on the ground, in other words by loss of memory.

This farmer had had this waggon for sixty years and still found it handy in the hay-field. His father had been a friend of Jacob Pike and both were buried in the churchyard of the great church at Edington where Alfred beat the Danes. I examined the waggon. The raves over the great hind-wheels had been lost and replaced by straight timbers, while the intricate under-carriage had been patched up by odd bits of unseasoned wood that brutally contrasted with the wonderful chamfering, hatching and carving of the original timbers. The former lettering and red and blue colouring had almost faded out. But the double shafts for the "thill" horse, the "vellies," the great hubs, the middle-staffs, the marvellous network of carved and perfectly fitting timbers of the under-carriage, there was enough of all this left to reveal what a work of art this waggon had once been. It was the most extraordinary mixture of beauty and shoddy, of solidity and flimsiness, of devotion and makeshift I think I have ever seen. The continuity had been broken and this yeoman had had the pain of propping up the old waggon like this because he could not afford to do anything else. His masters, the financiers and the politicians, had seen to that. When Jacob Pike had built that waggon, the country memory was still intact; when those spars and planks had been nailed on, the tradition was gone, and there stood the old waggon, suffering from loss of memory.

The Wisdom of the Fields (1945:61-2)

Hedging

THE spiritual wilderness of our time expressed itself in a physical one. In all directions, by-roads are being drilled, licked and hammered into imitations of urban highways, while the hedges on either side of them riot in disorder and mildly inquisitive cows thrust their heads through the gaps to blink at the passing show. Extremes meet.

Suddenly, this year, a partial change has appeared. The run of the abscess is being checked by the surgeon's knife and the stealthy insidious creep of the hedge out into the field on one side and the grassy border of the road on the other arrested. In February about a tenth, this month about a fifth of the sprawling hedgerows, over-luxuriant in the crown, lanky in the leg, rotting and gaping at bottom, have been taken in hand. All over the countryside, hedges that

"Like prisoners wildly overgrown with hair,
Put forth disorder'd twigs,"

I see being attacked with the billhook, or as Shakespeare would have said, being pleached. In *Much Ado,* the prince and Claudio walk in "a thick-pleach'd alley," and Hero, bidding her wench call Beatrice, says, "Bid her steal into the pleachèd bower." The word is, of course, older than Shakespeare's use of it, and I find it in Fitzherbert's *Boke of Husbandrie* (1534) – "To plasshe or pleche a hedge." The country word survives to-day in "plashed," "plushed" or laid-down hedges, but the art? Everywhere I see badly pleached hedges for the very good reason that the born and trained hedger is as scarce as the born and trained thatcher to meet the demands upon them. The miles of hedgerow in sluttish disarray call in vain for the qualified doctor to treat them. The amateur hedger is in his element, and many a hedge is so incompetently laid down that many of the "splushes," as the older countrymen of Bucks and Oxon call them, are doomed to perish.

There could hardly be a sharper object-lesson in the folly of allowing the traditional country crafts to die out. It takes a craftsman to pleach a hedge as it does to slat a house-roof, thatch a barn, plough a straight furrow, build a field-gate, or doctor a sheep. If their numbers are lessened by insecurity, bad pay, mechanical makeshifts, agricultural decline and the lure of the town, then the country has to begin all over again where it started from three, four or five hundred years ago.

This is precisely what is happening to the ancient art of hedging but with the saving clause that, thanks to the resisting power of the family in country life, to inheritance from father to son or from uncle to nephew, a scattering of knowledgeable hedgers does still exist to teach the young idea and the reclaimed hedge to shoot. My heart leaps up when I behold the blunderers hacking and slashing about with their billhooks, not out of superior knowingness – since the worst of them make a better job of it than I could – but because the sight of one of the useful arts in the act of being relearned and recovered is an exhilarating one. It is good for us to find out how little we know about fundamentals in comparison with the traditional knowledge of past ages, how great is our need of the master-craftsman business methods and machinery have dispossessed, how unreal is the theory (held in especial favour by professors) that the past can never return and catchpenny the phrase (uttered by what is called the man-in-the-street, which means the newspapers) that you can never put the clock back. Over thousands of acres of the countryside that salutary movement in reverse is taking place this very day. We are learning over again what our ancestors taught themselves far back in time – how to pleach a hedge.

Hedging in the traditional style has been subject, as lace-making was, to diffusion from a common centre and Mr. Hennell in *Change in the Farm* records how skilled missionaries were sent from Leicestershire into West Suffolk and from Northamptonshire (John Clare wrote an enchanting hedgers' ditty) into Shropshire. From signs in the hedges where marks of former pleaching may be seen, I gather that Bucks and Oxon as far as the limits of the stone country were suburbs of the capital of the hedging craft, which was the Midlands. This hypothesis is strengthened by the fact that a few hedging families (the Nappins of Oakley, for instance) survive in the region, and one day I was happy to find a newly pleached hedge in a lane leading to the remote village of Waterstock (where the Mummers still act a spavined version of the old play of death and resurrection). This hedge at once revealed the hand of the master and showed even the more plausible-looking hedges in the neighbourhood to be the effect of window-dressing rather than of craftsmanship. It is only by comparing a piece of authentic hedgemanship with what only masquerades as such that the fine points and niceties of the craft can be duly appreciated. I made enquiries and found that this very hedge had been entered for competition at the autumn show of the nearest market town and that its maker had won the first prize there for two or three years in succession. Three times the judges inspect the hedge, immediately after pleaching, again when the stools begin to shoot and once more when the hedge is in early leaf. The object of the last visit is not only to judge how well-made it is in feather as against its anatomy, but what is the proportion of dead to live wood in it. Your expert hedger takes care to drive in a certain proportion of stakes that will sprout. These stakes make contributions to the living hedge over and beyond acting as uprights for the "eddering" and securing the layers. One live stake will hold a hedge better than half a dozen dead ones, and, if the judges find less than one such every five yards along a given hedgerow, there will be no prize for the hedger.

This particular hedge, so neat and well-dressed, was enlivening enough to send me hunting for its author, and, when I had found him, I prevailed upon him to come and pleach a chain of my own hedges, which, being hidden behind wattled hurdles six feet high, had virtually ceased to be a hedge at all and had reverted to a discontinuous line of leggy hawthorns with swollen heads. He was young, which was a happy surprise, a cowman on a large farm, with bright black eyes set in a dolichocephalic head of unusual length and surmounted by a mop of black hair very much more unruly than any of the hedges he had pleached. Add to these a vivacious face, brisk resolute movements, smallish bones and a general elasticity of frame, and there stood before me a perfect specimen of Neolithic man, of old a champion waller, tomb-builder, flint-knapper and thatcher, who usually crops up in larger number further west and south-west. I ventured a shot in the dark, which from my past experience of craftsmen might the rather be called a shot in the twilight. "It was your father who taught you how to lay a hedge, and he was a prize-winner and so was his father before him. Isn't that true?" He looked at me as though I were some kind of magician, such as his far ancestors had been quite familiar with, and after that (the shot had found its mark) all was plain sailing.

Since my hedge was too far gone to depend entirely upon itself for rejuvenescence, a cartload of brushwood had to be stuffed into the wider gaps. This is laid down in the same manner as the live "plushes" or "splushes," the local vernacular for the older word, pleaches. But this brush is always laid along the hedge-bottom, so that the live wood above it may gradually enfold and finally replace it. The master-hedger, when laying down his plushes, never bends them down and breaks them. His object is to leave them attached to the parent stool by a slender "tongue" of wood, and to this end he chips or axes the place of fracture with light downward blows of his billhook until he has all but cut right through the wood, and this prevents any splintering or cracking when the stem is laid down. I have never seen an *ad hoc* hedger do this unless the wood is too thick to be pleached by any means but by the blade. As often as not, the amateur neglects to pleach as close to the stool as he conveniently can, to chop away the projecting butt of wood on the other side of the notch and to "tongue uphill" the section of the layer closest to the stool. The master-man never fails to perform all three desirables. He cuts close up to the stool in order to make sure of the current of sap passing from the stool to stem, or, as Fitzherbert put it in 1534, "for elles the sappe wyll not renne in to the toppe kyndily"; he removes the butt after the cut to encourage the new growth from the stool and prevent the rain from running down the stem into the heart of the wood, and he leaves the stem and stool attached by as slender a lath of wood as possible in order to guarantee freedom of growth. Skilled tongueing is always the mark of a good hedger, because, the thinner the tongue, the less does the layer take the sap from the new shoots springing from the stool. Though his aim is to lay his plushes in diagonal lines as near parallel as he can get them, he knows that they are of secondary importance to the stool. That is why the first thing a judge looks at is the bottom of a hedge. Well and truly laid, it will support the hedge even when, in late summer, the weight of the leaves is heaviest upon it.

Even when the hedge is "gappy," the trained hedger makes a careful selection which of the plushes to lay down and which to cut right out, and he chops away the stems he does not want close down to the butt of the stool. By such means he is able to control the evenness of the plushes, so that the hedge will not grow up too "bunchy" in one part and too thin in another. Like all true craftsmen he makes use of three eyes, two for the job immediately in hand and the third or mental eye for the general effect of a series of jobs. On the side where the hedger is working, the hedge presents a curiously shorn appearance, and this is because all the side-branches between him and the stem to be pleached are cut away, partly to give plenty of air and light to the stools and partly so that the layers may lie aslant along the hedge in more or less parallel lines. He judges with his eye where the stems will fall. The bushy tops are laid along the other side of the hedge, and, as he moves along a given length (the work is done in sections), he drives in stakes a yard or so apart on each side of the layers in order to hold them fast. The thicker the bole of the stem to be pleached – and my hedger told me that he thought nothing of laying down a tree "as thick round as my body" – the more ingenuity he has to exercise in order to edge it in line with the other plushes and push its head across to the further side of the hedge without snapping the tongue. I needed no further testimony to the high degree of skill exerted by a master in the craft than the sight of my own hedge that, with more gaps than growth before treatment, presented a long line of diagonal and well-nigh equidistant plushes when it was finished. It looked as though nothing bigger than a hedgehog could get through by the time that my hedger had taken off his mittens, his apron, his pair of knee-leathers tied on with string and the old coat he had donned for the occasion.

As the pleaching continues section by section, the number of the stakes is multiplied until there is only a foot or little more between them. All are sliced by an "uphill" blow with the billhook after they have been "sharped" and trimmed by a succession of swift, left-handed, back strokes. Whitehorn is the best wood for staking (Fitzherbert's *Bake of Husbandrie,* 1534, says oak, crabtree, elder or blackthorn) because it is superior to all woods in it powers of sprouting and willow of the third or fourth year is the second best. The last process is the "heathering," called "eddering" in Northamptonshire, or the plaiting together of long shoots of bramble, willow or hazel between the stakes. "Eddering," used by Fitzherbert in 1534, is probably the older term of the two. Bramble is the best for the purpose, and hazel or willow of the first or second year are alternatives. My young hedger cut both heathers and stakes from the old willows clustered on the banks of the stream where two pack-horse bridges crossed it. Most hedgers that you see about nowadays get their stakes out of the hedge they happen to be laying, and I doubt whether they know that one wood is preferable to another for durability or that a slender wand that is alive is better than a stout stave that is dead. In the neighbourhood of Silchester, the staking and heathering are subject to a local technique. Thick stakes are driven in while the plushes are being laid down and, when the heathering is due, slenderer stakes are fixed in the ground between them, the completed row making a closer fence than in the Midlands. The withies are then *wattled* to a depth of from six to eight inches, so that the top of the new-pleached hedge has the appearance of a wattled hurdle. This makes, of course, for the strongest support possible for the growing hedge.

In the Midlands, the first rod of the heathering process is placed with its butt against the inside of the stake, the second against the outside of the next stake, each rod being twisted round the other so that a double cable is in the end stretched along the whole line of the hedge, firmly held in place by encircling each stake on either side. The chief concern of the hedger by training and inheritance is to secure a taut and straight line of heathering and for this purpose he drives in the stakes with the back of his axe. In the old days, he used to do so with a hedge-maul, a club of hard wood. I lent mine from my museum to the young hedger – an ancient and battered club of elm-wood such as the natives of Cerne Abbas might have used 1,500 years ago in driving St. Augustine and his missionaries out of the village sacred to the Cerne Helith, with *his* gigantic club carved on the hillside. He was enchanted with it, but it was too heavy for him. For a hedge long neglected (twenty years is long enough between one pleaching and another), four lines of heathering make a solid workmanlike job and there should never be less than two.

Neatness and symmetry are obvious credentials of a well-pleached hedge, but the experienced eye of the countryman distinguishes the craftsmanly from the indifferent hedge by further "observables." He will look for a firm bottom, whether the cuts are clean and made "uphill" with a sharp billhook and if they are made as close up to the stool as possible. He likes to see the plushes laid well off the stool, the hedge evened to the same thickness throughout, greedy or strangling growths like ivy, elder and ash cut away or very severely pruned, the ditch cleaned out and the "ditchings" laid as a bandage round the stools. Straight firm heathering is a cardinal point and that means stakes from a foot to sixteen inches apart (Fitzherbert allowed 2½ feet in 1534), with a live one every five yards, and each stake lopped at the top to make a level line. He will be particular about clean and frequent "tongues" from the stool along the layer and he will examine the stools themselves to see whether the projecting pieces are struck off "uphill." Are the plushes laid down in more or less the same line of accidence and have the side-branches been cut off close to the stem on the side where the hedger stands? These may be called, Tusser-fashion, fifteen good points of hedge-husbandry. The best hedge for laying is either whitethorn or blackthorn, preferably the first because the stem-growth is straight while the branches make for thickness. Crab makes a first-rate hedge, but the tangled heads make it hard to lay. Wych-elm is excellent, but the cattle strip the bark. The wood of maple is too brittle for good laying. The soundest qualification of a good hedger is that, stored with inherited or half instinctive knowledge, he should adapt himself to circumstances and the nature of the material.

The Sweet of the Year (1939:20-31)

Husbandry and Work

WHAT with mowing-machines, swath-turners, tedders, side-delivery rakes, collecting-rakes, motor elevators, loaders and stackers, the mechanized army of June that standardizes hay-making and turns it into so much task-work in meadow after meadow from the toe of Wessex to the tip-top of Northumbria, leaves me next to nothing to say about the once ritual of the Cotswold hay-fields. These lumbering saurians of Progress accomplish one thing and one only – they make a quicker job of getting the hay off the field. But at a cost not only to the pride, the aesthetic skill and the companionship given rhythmic expression by the practice of the old rites and yearly remembrances of the haymakers, but to the quality of the hay itself. The machine pinches off the hay; the scythe cuts it and so encourages a clean and brisk regrowth. "There's a fortnight's difference between a good crop and a bad 'un," a farmer said to me. But what does the crop matter if the fortnight can be saved? The deteriorative effect of the machines is best seen in the impoverished eastern grasslands of the Oxford Clay bordering the hills. Partly through lack of labour and partly lack of any respect for what the earth produces for them, either the haymakers or the mechanical tedder, which can only turn the hay on an even surface, scamp the tedding and drying of the hay. The boys or old men pile it on the rick still half green. This, of course, renders the unthatched rick not only shapeless from the sinking of the green stuff but liable to take fire. To obviate the one they "put a policeman" round the rick by propping it; to prevent the other, a hole is cut through the middle of the rick from top to bottom and hurdles are inserted to let the heat out. What do the old-timers think of such shoddy opportunism? But the arable-lands of the hills do check such rankness from developing there.

No doubt scything the meadows was terrible hard work in the sun, and after playing Andrew Marvell's mower, Damon, myself, only for an hour with the scythe, I no longer marvelled at the quantities of scrumpy consumed among the hay-fields of the past. But mowing compels a rhythm, and, if the line of least resistance be replaced by muscular effort; if the scythe slashes; if the stroke is hurried and the body tenses; fatigue will soon correct the error and restore the even sway and poise from the hips. The machine stupefies the senses, but the heady scent of the falling grasses, the intimate contact with the earth, the very hiss of the blade quicken them like wine. It was surely by such means, by guiding and not wielding the scythe, that the stalwart countrymen of thc past were enabled to begin mowing from 4 a.m., when the dew on the grass made for a crisper cut, till 6.0, continue after breakfast until 10.0 and again till dinner at 12.30 p.m. or 1.0, take up the burden again to 4.30 or 5.0 p.m. and continue with only one more break until 10.0. In those days, too, the "leader" of a "flight" of mowers set the pace. Fifty years ago, these mowers were still a long way off the Elizium of the Eight Hours Day. Having gained it, we have ignored the significance of what George Sturt said in *The Wheelwright's Shop* (1923): "It should be pointed out that in those days a man's work, though more laborious to his muscles, was not nearly so exhausting yet tedious as machinery and 'speeding-up' have since made it for his mind and temper. 'Eight Hours' to-day is less interesting and probably more toilsome than 'twelve-hours' then."

Shepherd's Country (1938:106-7)

The Christmas Tree

THE Christmas Tree was one of my lives. It is an old beer-house and the club of the agricultural labourer. Others frequented it, a fraction of the landowning class, local officials and ex-policemen, a dusting of "gents" who lived in the village or had week-end cottages, and occasional aliens. These, all except the last named, mixed well with the authentic native stock and were consistently welcomed to it with that gentle courtesy which I have always found to be the inherent quality of the landless labourer in his intercourse with worlds not his own.

I have pushed open the doors of the Christmas Tree about seven hundred times and never met with that forbidding or clannish withdrawal into themselves which is related of other workers on the land in other countries. To the alien they are reserved but not out of any half-resentful stiffness at his presence in their midst. They are merely shy and rather embarrassed with what is unfamiliar, but boorish, prickly or uncouth never. There is pathos in their reception of strangers, for they never betray in word nor manner that the Christmas Tree is their own, their own by right of custom, of daily companionship and of something intangible and indescribable, an effluence, a sentiment which is felt if not seen in the smoke rising from their pipes, in the smoke of the fire, in the air you breathe, in the very grain of the wooden benches and the bubbles of the brown liquor . . .

It is on the same plane as is my bond with the labourers of the Christmas Tree, except that I also share with them a common humanity. A common humanity, but do I feel that with other men who meet and talk with me? Certainly not – the points of contact are those of like tastes, personal affection or intellectual interests. It no more occurs to me to see them as men risen from the green body of nature and breathing the universal air with me, sweeping the vision across the same stars and warmed by the sun that shines on all alike than it occurs to me to be conscious of my tongue that speaks with them or my eyes that behold them. But this sentiment, this soft, impersonal but musical emotion is always present when I am sitting in the bar-room of the Christmas Tree. It is, I think, the reason why I can sit among them with my pint and feel neither constraint nor restlessness when, as often happens, no word is spoken between us. As in no other company I know or have known, they possess the mysterious faculty of giving out a sense, almost a radiance, of composure which extends beyond the four walls of the inn and is mingled with the tranquil immensities of the night without. We are not an enclosed and lighted space cut off from the sleeping green earth and the air that enfolds it, but a spark struck out of those elemental harmonies.

I suppose that another reason why our mutual heritage of a common humanity quietly possesses me when I am with them is because they are nearer the primitive than other types and classes of men who speak the same language but not in the same idiom as they do. They have a stake not in the country but in its soil. In an odd way, I feel that they are innocent; the blood of Abel is not on their heads nor theirs the responsibility for the fury and the fraudulence of the civilised world. It was one of them who plucked the apple of knowledge because at that time all men were as they are, except that they had no masters. But when he had plucked and eaten and brought into the world all our woe, all our quests and adventures of the mind, he ceased to be one of them and was cut apart from them for ever after.

They are closer to the primordial men who were content to live on the earth and all it bore and all whose hoof or pad ran swiftly over it before some schemer or ideologue was smitten with the thought of men living on their fellows. The quietude they diffuse thus partly emanates from a nearly changeless, an immemorial tradition of a mankind in correspondence with the earth, its seasons and the alternations of its day and night, which labours and puts aside its labour. I do not say that this innocence which sits so well and unobtrusively upon them is the fruit of any mystical communion between the soul of man and the beauty of the field. It is because to own, to plot, to covet, to drink the maddening draught of power has not been given to them any more than to see visions, to pursue the witch of beauty nor to map the universe. But their traditional inheritance of the primal nature of men, the uncomplicated conditions of their lives and intercourse with the first and last realities have invested them with a dignity which breathes a concord and so a beauty all their own and reaches to the very clothes they wear. They are clothes which, if in halls an excrescence, are under the wide heavens as fit, as integral a portion of the scene as the cherry tree in blossom.

A little of them I know, not only because of the countless sessions in their company, silent or rippled with talk, but fragmentarily my labour has been as their labour. I have not laboured for a master, as they have, nor with their skill and rhythm and continuity, nor has my bread been to win from spade or plough or billhook. They are versed in the manners and customs of the earth as I am not. But like a child who daubs from his father's paint-box, I have scratched the sod and made it bear as they have. I know what it is to wrestle with the land in a preoccupation that lulls the questing imp within and delivers the whole being into the security of the universal tempo.

This and to sit with them and drink with them night after night gave me a rough pass into the kingdom not of their minds nor souls nor emotions but of their being or, should I say? of a diffused being they share in common. They were as different in appearance and personality as in age but there was an elemental commonalty between them as between leaf and grass and wayside flower. By its aid I could look dimly through them and backwards into a hinterland nebulous as the fields of Dido's ghostly flittings and see their antique fellows who millenium after millenium took their arduous joy of the day but no thought for the morrow, who delighted in mouth and eye and limb which gave them the riches of the senses and in their caves slept without evil dreams. I was always safe with these men: if their brows were lined with weather rather than thought, what lay behind them was serene and without guile . . . We had lots to talk about at the Christmas Tree all the same. It was something of an agricultural college for me, and through it I acquired a smattering of most palatable erudition. In my rawest days as a gardener I made liberal use of a substance called Somebody's Vine Manure. I scattered it about like manna until I learned how it drew the nitrogen out of the soil and in the end impoverished it. It was the Christmas Tree which first taught me the same lesson about artificial manures as a wise doctor gives his patient about sleeping draughts. William, shrewd old James's very personable son with his fine sensitive features and delicately-poised physique, was one of my professors. There was a night when I sat between them and we spent an hour and a half discussing the ways of moles whose increase in the dryest of dry years had created a problem for me in my new garden. What William said was capped by James and James's words were chained to William's, so that like a clockwork toy I was perpetually turning my head from father to son and from son to father. "I expects," James closed the stream of information, "moles was sent to do their job same as other people." I owe it to William that I shall never be so *gauche* again as to confuse broccoli with cauliflower. Spring cauliflowers are broccoli, and only cauliflowers are cauliflowers. Broccoli must be cut at just the right time: if not, it will change and flap open and go yellow within twenty-four hours. William plucks a leaf from his broccoli and screens it from the sun. We would debate at length the relative values of the french bean and scarlet runner, nor were brussels sprouts, carrots, peas and onions forgotten. At times, the talk was so technical that I went out no wiser than I came in, but I usually managed to pick up, as the robin did the crumbs from our table on the lawn, oddments of information about the tricksiness of onions, the virtues of early Pilot (but you have to be very careful about plucking the pods lest the plant yellow) and late Ne Plus Ultra peas and similar good things. It was the Christmas Tree which shook my loyalty to Thomas Laxton as a pea prodigy. Sometimes, these discussions became hotted up to the point when stems of pipes were pointed at the unconvinced. I remember one such when the unanimity that if you top a Brussels sprout before the babies are formed close to the earth it will be a poor cropper went crash upon the vexed question as to whether or no runners should be nipped in the bud. When doctors fall out, what is the student to do?

Novice as I was, I never made the mistake of listening with only one ear. I was at school again with the difference that I was now learning things of fundamental importance to the life of the community. To think that I have lived more than forty years without recognising a Pilot pea when I see it and that thirty years ago I knew how many ships fought at the Battle of Arginusae! The mercy, the self-protection of forgetfulness! But the manners and customs of the vegetable kingdom was only one of our themes. Many an evening was crowded with variety of converse and after one of them I jotted down the topics of an hour – crops, weather, roads, distances and the qualities of beer. On that same evening I learned that the local name for wild fritillaries was "frorechaps," and there is a "Frorechap Sunday" in April when the beauties are ravished. On another, I was piled with exhaustive information on the subject of finding water and sinking wells, my ignorance of which had been, previously to that evening, without a flaw. I was amazed at how much these men knew about water-divining, its theory, powers and limitations. Another was occupied with the question of how best to start up a car in cold weather. Though the association of crops and cows with motor-cars appears indirect, these herdsmen and ploughmen knew as much and more about cars than the average mechanic at a garage or petrol station.

Country (1934:134-5 and 140-5)

English Cheeses

SOME weeks ago I stayed the night in a noisy but faded little market town, and, while dressing in the morning, happened to glance at the yew-tree, whose branches darkened the window. It was full of crossbills. I called out to my wife: "Do you want to see a sight as rare as eating a good English cheese?"

The decline of the English cheese is one of the best examples of how commercial values have ruined good taste and discrimination. In 1850 it would have been difficult to find a single farmhouse in Britain which did not make its own cheeses, just as it cured its own hams, baked its own bread, and brewed its own beer. People at that date used to talk about "good cheese hands," as we to-day talk about "green fingers." In still earlier days, most of the farmhouses possessed cheese presses, some with a great squared slab of stone, others with a square wooden box filled with stones in which the cheese was squeezed, on the same principle as cider in the cider press. Where the cheese press was absent, the cheese was wrapped in a cloth and pressed down into the vat under a heavy stone. Weyhill cheeses – has England forgotten Weyhill Fair, where the cheeses of the west met the sheep flocks of the east? Seeing what Weyhill is to-day – a fringe of vulgar hoardings round a green – it is no wonder that this illustrious mart of English cheeses is forgotten. Where is the Mardale now by Hawswater with its old church, its old inn and its Wensleydale cheeses of uncommon excellence? Buried under the water like Lyonesse to make a reservoir for Manchester. For *Manchester!*

How many out of this multitude of local cheeses survive in our day? English cheeses have vanished with English cooking, that once had none superior to it in Europe, even in France which, if first in some things – soups, for instance – was second to us in as many others. Now English cooking is derided as the worst in Europe, and we will not even call an English dish by an English name. We not only live on tinned importations but eat vamped-up messes under French or other foreign names in the fond delusion that we are being Continental. The menu has generated as bastard a brood of monstrosities in language as any spawned by scientific invention. And where are our cheeses? All, all are gone, the old familiar slices, except Cheddar and Stilton. I once bought a pound of Cheddar at Cheddar and left it at Cheddar (in the ditch), and the fact is that properly matured Cheddar is rare. It too often has that venomous bite to it, that rancid comeback that betrays how its makers have forgotten the cardinal desideratum of all good cheese (as of most other things) – the ripeness is all. That depends upon the right amount of pressure exercised by the hand-press, the condition of the curd and other factors known only to the delicate touch and the natural aptitude that usually come from inherited knowledge. Besides, most people happily eat Canadian Cheddar; they do not know the difference, just as they do not recognise the difference between a good and an indifferent cheese. So, a genuinely ripened Stilton is as rare as a needle in a bottle of hay.

A handful of other English cheeses most precariously survives – Wensleydale, Cheshire, Dorset Blue Vinney, Leigh, Leicester, Double Cottenham and Gloucester. Most people have never heard of them, and yet each is as fine a cheese as exists anywhere in the world. Personally speaking, I prefer Double Gloucester to any of your Dutch and Swiss cheeses; for richness and softness of flavour matching its beautiful colour of autumn beech leaves it is without a peer. There are, I believe, two shops in the whole of England where you can buy it – the proud Englishman prefers pieces of foreign stuff like soap, wrapped up in silver paper. It is not so strong as Wensleydale and Leicester, fine cheeses both, but it has a fullness and a tone that make it the perfect cheese with celery. *David Copperfield* is our only authority for what it used to cost. The baleful Mr. Murdstone in his lesson to David says: "If I go into a cheesemonger's shop, and buy five thousand Double Gloucester cheeses at fourpence-halfpenny each —"

Quite recently an Oxford provisions dealer had a felicitous idea. He organised a cheese week, so that English people for the first time in their lives had the opportunity of choosing between more than two English cheeses. As a result of that week I learn that Double Gloucester is to be on the permanent list. By the shade of Cobbett, that is what I call good news. It will help to regenerate the decadent English taste in food. For half-a-dozen farms in the pastures of the Vale of Berkeley it will keep the wolf from the door. Best of all, it will perhaps save an English rural industry from extinction. How remarkable it will be for English people to discover that the English countryside can produce good things; that it is something more than a series of sites for bungalows, aerodromes, garages and roadhouses!

A Countryman's Journal (1939:122-5)

The Barbarian Invasion

BETWEEN Totternhoe and the ridge of the Downs above the Icknield Way, the land, which is quite open, takes a fine crescent. Rows of houses have been built and are being built on its slope and in such a way that they completely destroy the linear continuity of the sweep. There were miles of open plateau for these houses from which to choose an inoffensive site. But no, they must choose the one position which ruins the composition of the whole landscape. Some of these houses are of the usual makeshift, pretentious, inorganic type; others are of a not unworthy design. But the latter represent just as gross a violation of the scene as the former, simply because their sites have been selected where there ought to be no houses at all and without any relation whatever to their environment.

The things I saw did not compass the hundredth part of a devastation flung far and wide over England. Even the little towns help to spread it. They throw out their feelers just as the big towns throw out their suburbs. First, the arterial road; next the houses along its rims clinging like green fly to a stem; then tributary roads boring into the adjacent country with their "desirable frontages." Or the towns send out whole colonies like Peacehaven, as Crete and Egypt and Sumer colonised the lands east and west and north of them. Or the remoter districts are peppered with bubukles and whelks called modern houses and set up, as in the South Cotswolds, on all the points of vantage.

What does it all mean? It is the effect of a great migration or exodus from the towns into the country, and this movement is just as significant and symptomatic as the tidal drift a century ago in the reverse direction. The Industrial Revolution was responsible for Exodus I; the fruits of the new system or new chaos for Exodus II. The Industrial System drove the country people into the towns; its child, the Machine Age, is driving them out into the country again. The first movement was on the whole more forced than voluntary; the second is more voluntary than forced – but in each of them both elements are present. Social and economic causes (those of indirect coercion) play their part in the contemporary movement, but I believe that its emotional and psychological causes run deeper. The country itself has suffered both ways. The first polarity sucked up its crafts, its interests, its agriculture and the life of its local communities. The second, and expulsive influence, robs it of its beauty and its peace, and discharges upon it an urban horde whose ignorance of rural conditions is without a flaw. This reverse or centrifugal movement has no more deliberation nor intelligent purpose in its urgency than a river in spate, and all that the flood intends is that its waters may cease from flowing and be at rest. It is a flow of escape, and it lacks the discipline, the fixed orientation and the predatory destination of the old barbarian waves of exodus, the Kassites into Babylonia, the Tai Shan into southern Asia and the Dorians into Greece. But a foreign invasion it is, as their invasions were, because it is the town overflowing into the country, and there is as little understanding or knowledge of the one by the other as existed between those that dwelt in tents and the Cities of the Plain. I speak as historian, not as moralist: am I not one of the horde myself?

The consequences of a breaking out thus blind and chaotic, of a driftage which has only momentum to take it along, affect the townsman in one direction and the countryman in another. In the areas thickly settled by the invaders, the native population is driven out. The farm, the village pub, smithy and shop, the barn, the sheep-fold, the little market town, all vanish. The original stock is displaced. Even where this stock still clings in diminished numbers to its native village, its standard of living is much lowered. The urban immigrant buys up the labourers' cottages which come into the market and reconditions them. This process drives the labourer into cottages not worth the spending of money to refit them. In other words, he is squeezed into a hovel. Landlords who own the better class of labourers' cottages are, the less conscientious among them, always looking out for the town-buyer, and this makes their tenancies the more precarious. In districts where the up-rooting is less drastic, the manners, the psychology, and, what is still more vital, the traditional associations of the rustic with the land of his plough or spade or billhook or milking-stool, are thrown off their equilibrium. A mongrel dimorphism between town and country takes their place, and the townified countryman is many degrees worse than the countrified townsman.

The townsman himself is suspended in a vacuum. That is one element in which his migration differs from those of the Tartar or the Puritan father or other peoples, sects or tribes of the past. It differs because he does not cast off what he leaves behind. He still clings to it for his livelihood, his social environment, amenities and recreation, and also, where it exists, for his intellectual life. He has little or no commerce with the country itself. The country is where he lives, be it in temporary or permanent residence. He lives in the country, but he remains as much the townsman there as John Clare remained the countryman when he visited the literary circles of London. In the town his heart is in the country; in the country his associations are with the town. He repudiates the town but without adopting the country. Actually, the process goes even further than that. A large percentage of the rusticating townsmen makes use of the country, except for week-ends, merely as a dormitory. Every morning except one, the holy-day, the town sucks them back again. The bits of paper fly into the air-pressure shaft. I do not say they can help themselves: that is the frightening part of the whole business. The mouse scuttles away, but the cat's paw is always raised to drag it back. Thus emigration, or what Professor Toynbee calls the Völkerwanderung into the centres of past civilisations, bears a partial resemblance only to the emigration of contemporary population from the towns to the country. The flood-movements, new and old, are alike in their volume, their destructive effects and the dislocation they cause among the peoples they invade. Leaving out secondary unlikenesses, they are dissimilar in their purposes and their impetus. It might be said that, as the Völkerwanderung takes the goods and the cattle with them, so the migrants of to-day pack up the towns and transport them into the country. But the parallel breaks down because, like Lot's wife, the town-migrant looks behind him, and that the barbarians never did. His state is one of oscillation between town and country. He is Mr Facing-Both-Ways.

Still more radical is the opposition of motive. The pastoral nomad seeks the town-settlements for plunder, while the townsman is moved by love. That love emanates both from a negative reaction and a positive impulse buried within his subconscious. If he does not consciously recognise that our industrial and mechanical cultus has failed to bring him happiness or peace or contentment or zest in living or the sense that what he does is worth doing, he acts as though he were perfectly aware of it, and that is certainly the true source of his action. Pressure of population is only a subordinate reason. Drift on such a scale has a deeper and psychological derivation and the strings that pull him are drawn tight by unconscious ancestral memory. Up to the end of the first quarter of the nineteenth century, the town lived next door to the country; here was the town and there was the country, both intact and fitting as close to one another as head to trunk. A town was inset into the country like a nugget into its matrix. The vast ramifications of modern city life were non-existent and when the "land" of England or France was spoken of, that was what was meant. England meant, not London, Manchester or Leeds, but the English land, and this is not the less true because people did not travel so far nor so frequently nor by mechanically propelled transport. They did not have so far to go in order to get to the land. It is not the less true because a diffused sentiment about nature and the land is comparatively recent in England. That common sentiment arose when living on or by or with the land was ceasing to be an axiom of life. It was detachment which created it and a poet like John Clare, the organic countryman who is self-conscious about the country, is an excessively rare phenomenon. It is this blood-attachment to the land which, himself unaware of it and as little knowledgeable of it as a mole of geological terminology, draws the townsman back to his ancient home.

It is love which moves him. On the last Sunday in April, which is "Frorechap Sunday" in these parts, I travelled a few miles from my home to a meadow "quite o'er canopied" every spring with the wild fritillary. At this time of the year, it is or rather used to be a mottled sea of fritillaries. Over the entire pasture of many acres I found two. All the rest had been plucked or uprooted. Love had got there before me. As the ivy loves the tree, so did visitors from the town love the upland pastures of a Wiltshire farmer who broadcasted the consequences of their affections – his gates left open for the cattle to get out, his trees mutilated, broken glass, paper, tins and worse strewing his fields, soap and petrol mixed with the water in his troughs and one of his ricks set on fire.

These and phenomena allied to them are only the by-products – or should they be called the by-blows? – of the town-dwellers' passion for green England. His more respectable devotion (even though that, since the town only lets him out at the end of a string, is a kind of bigamy) is expressed in his settlements or camps in the conquered country. The camp-followers brought up the rear of the older type of barbarian invaders. To-day, they go before the van of the country-loving host. They are the speculative builders, and their share in the paradox is to destroy the country, the craving for which has set the multitude upon the march. *E pur si muove.* Onward,

Christians soldiers! That there is no country left where the migrants settle, makes no difference neither to the quest itself nor to the satisfaction which attends its goal. The flower is plucked and lies dead in the hand. But to its possessor it is still the flower that lives for ever and at last is his. Mr Chesterton, I think, has called attention to the paradox of the country-lover killing the thing he adores. But what seems to me a still more curious reflection upon the mentality of our city-bred age is that the vast majority of the new settlers are totally unaware of the fact that their notion of living in the country is an illusion. That which they sought, which moved them out into a novel environment and for whose sake they taxed their purses and broke their habits, has dissolved like the fabric of a dream. What most of them have got is a bad imitation of their urban setting, houses to the left of them, houses to the right of them, all the spit of their own, rows of shops, perhaps Woolworth's and a cinema, fenced back-yards, and a view over to the next suitable building plot. But everything is all right: they are living in the country.

This extraordinary myopia is what industrialism and mechanisation between them have done for the town, they, or rather the social system which created and was created by them. What else could be expected of a system that loses the end in the means, that makes an artificial scarcity in the midst of plenty, that imposes poverty and uniformity upon the free human spirit, depopulates the country and over-populates the towns, puts millions of men out of their jobs, lives by strife and in its collapse is driving all Europe into suicide. It infects the whole world, its enemies no less than its apologists and those who profit by it. It is one of the saddest ironies of our contemporary life that the host which flees from it to the harbourage of universal nature should carry that spiritual infection with it. The people who built our cathedrals and parish churches, our barns, cottages and manor-houses, had far less feeling for the face of the land they made more beautiful than our latter-day migrants who deface it with the pimples in which they live. A devil has entered in upon them which we call to-day vulgarity and blindness to values of truth and beauty (though not, I think, to goodness), and the industrial system has destroyed their resistance to its passage.

But the love is there, it is a reality, a love of beauty and the nature which bore them, but which they have been deprived of the power to cherish and understand. Their omnipresent gardens bear witness both to this love and their weakness in making something of it.

Compare their gardens with the average cottage garden of the countryman, and you can see at once that they are devoid, ninety per cent. of them, of that instinctive rightness and perception that bless a patch often no larger than the cottage kitchen. This, therefore, is the problem that confronts us. It is an airy nothing to propose stemming such an invasion: as soon set up a fence of wattles against the hunger of the tide. Education?...The only way to educate the invaders is to educate them out of the education they have received. What is needed is that the invaders shall somehow get back what they have lost, a sense of right values. The only thing that can give it back to them is that very sentiment for the country lacking, nine times out of ten, in the people who could not do a thing aesthetically wrong in making the England of the past. The sentiment will only cease to be destructive, that is to say, when it sinks into the subconscious and this cannot happen until the false values imposed by the Machine Age are sloughed off. No social and economic changes can accomplish this because it is a thing of the spirit. They may prepare the way for it, but it can only come into being by the growth of a new vision born of the bitter experience with which those false values are already feeding us. If the European society of which we are a part falls to pieces by war, the survivors will have to get back a sense of right values.

Through the Wilderness (1935:110-17)

The Week-End Cottage

THE victimisation of the country by the towns has a bitter history behind it, dating from the times when the early Tudor courtiers and merchant adventurers speculated in the lands of the village communities for the purpose of converting them (and evicting the natives) into extensive sheep-walks for profiteering in the wool trade. The latest expression of this hoary rendering of "the devil take the hindmost" is also one of exploitation and eviction. Then it was the land, now it is the country cottage that has grown out of the land. The townsman is weary of the town and yet cleaves to it for his livelihood. So he solves the problem by taking a "week-end cottage" in a "picturesque" old village. Since demolition, road-making, dilapidation and lack of means or lack of interest in remedying it, these and deeper social causes behind them, have greatly decreased the number of "picturesque" cottages available to this disillusioned townsman, he cannot pick his cottage as he can a ripe blackberry from the hedge. Result – there has grown up in all country districts a new type of speculator, who makes farm- or cottage-hunting his business. He buys up all the farms and cottages he can (and here a depressed agriculture is an asset on his behalf), whether vacant or occupied, condemned by the local authority or not, and proceeds to "recondition" or modernise or Tudorise his finds, raise their rents three, four or ten times what they were, let them to the questing townsman and depart elsewhere, leaving a little income for himself behind him.

The villager cannot pay the inflated rent and out he goes, either to leave the village altogether, and so depreciate the reality of village life one step further, or to take a council house in the neighbourhood. Many council houses are built simply to meet this new demand on the part of virtually evicted householders for a roof over their heads. The necessary condition of these houses is that they shall be cheap, and so, not only do they involve a Government subsidy and a charge on the local rates, but they are nearly always a shattering discord in the harmony of, and a skin eruption upon the face of the regional and traditional village architecture. Thus one seemly cottage in a village is replaced by a masquerader in sham Tudor costume and an urchin of a council house.

I have personal knowledge of many such heartless evictions of field labourers, of many an honest cottage turned into a Tudor pretender and of productive gardens into playgrounds for stone gnomes and animal monstrosities. But it would be a mistake to assume that these speculators are always professional business men, one of which in every hundred possesses either architectural knowledge or aesthetic values. Many of them are residents in a village who have established themselves in it from outside, and are not at all what in 1577 old Harrison of the *Description of England* called "bruggers" or middlemen. Hearing the rustle of Bradburys in the trees as they take their morning strolls, they turn speculators. I know of one village, the two ends of which have been bought *en bloc* in this manner, the former residents (all native villagers) being dispossessed, the rents jumped up to as high a figure as a pound or more a week, and the new occupants all week-enders. I know of another village which by such means has lost nearly its entire stock of natives. The indifference to the welfare of the villagers is shocking, but is due not to hardness of heart but to a curious blindness to the fact of their existence. But these trespassers do, of course, include a certain percentage of people who know the difference between a good-looking house and a bad one. Sometimes, therefore, they will restore a dilapidated cottage with taste and discretion. But there is no doubt at all that the general effect of this traffic is destructive both to English villages and English village life. It raises the general level of rents; it mars the traditional beauty of the English village; it behaves like a conqueror towards the native population, and it substitutes something callow and unreal – transported urban culture – for something authentically rural. Lastly, a wedge is driven into the corporate life of the village. To the week-ender, the village itself is no more than a *pied-à-terre.* The better type gives subscriptions; the worser noisy parties, and infests the village inn to the discomfort and occasionally the demoralisation of the habituals.

But even the best type often exhibits so dense an ignorance of country life that a new division is added to the normal distinctions in the social hierarchy of village life – that between native and foreigner. Since, again, the number of week-enders actually exceeds that of the indigenous folk in some of the smaller villages, many a retired corner of genuine country England is being transformed from a village into a suburb. New roads appear for the Monday morning exodus; garages spring up and the tide of townish wants creeps in. The sons of husbandry also are magnetised into the towns, but they do not return for the week-end. In twenty years, a new culture has taken the place of the old in the village. But a village culture it is no more.

A Countryman's Journal (1939:41-4)

Conquered Country

OUR poor country! Hardly a day passes but some new indignity to its spirit, some fresh wound to its scarred body, come to my ears. Yet how could it be otherwise? There are two peoples in England, the town and the country, the conquerors and the conquered, and this division, made by the Enclosures of the eighteenth century in the first instance and the Industrial Revolution in the second, is, though little recognised, a far deeper cleavage and one affecting more vital issues than any between political parties. If the events of the last twenty years were tabulated into a convenient digest, the treatment of the country by the town would impartially appear as not at all unlike that of a foreign despot in possession.

The building speculator appropriates our inheritance and diminishes its food-bearing capacities. The county councils deprive the parish councils of their local powers and administer huge tracts of country like Persian satraps, or, worse, like a petty bureaucracy. The Forestry Commission with a dingy foreign conifer for its lance tramples underfoot our native scene and leaves behind it devastated areas tenanted only by gloomy spruces. The Ministry of Transport turns country roads into imitation racing tracks, while the thistles on either side of them, left to their own sweet vagaries by the human depopulation of our fields, crowd together to see so strange a sight. Urban mass-production stamps out country crafts, country food, country ways and countrymen. The mechanisation of agriculture which in no way increases the average yield of crops and whose only object is to substitute engines for men, is a form of industrialising and so urbanising the country. Read your urban philanthropists who talk about the country as though it were merely the recreation of the town, and what are their words but the voice of the conqueror? There are two Englands, and the one lies prostrate under the heel of the other, its drudge, and its means of enjoyment. Yet the oppressor, at last aware that the town in the long run depends upon the country, talks of putting the country on its feet again. Then let him take his heel off the neck of our immemorial own native land!

I am urged to these distressful accents by a letter from the beautiful little corn-village of Finchingfield in Essex. Under the Board of Education scheme for centralising education in the towns, the children of Finchingfield and many other villages in the neighbourhood up to a certain age are now taken off to Braintree, eleven miles away, for their schooling. This is now happening every where. "One more nail," as my correspondent writes, "is driven into the coffin of rural integrity." In other words, busy urbanising everything else in the country, the towns have now begun to urbanise its children. There is nothing like catching them young. In one generation, the Hampshire child will be untaught the stored country wisdom of a thousand years, the Devonshire child will hanker after the delights of Suburbia, the East Anglian child will not recognise the difference between a hawk and a handsaw and the Dorset child will regard the tongue of William Barnes and his own grandparents as a foreign language. The diversities of a home-made speech are already being replaced by a universal Cockney. The old village school had its faults, but at least it did, whether unconsciously or by intent, teach its charges something about the place where they lived. At its least, it in no way robbed that inherited fund of what George Bourne in *Lucy Bettesworth* called the "primitive knowledge" of the countryman. That was the knowledge, expressed in ten thousand different forms, which redeemed England from wilderness. Modern commerce is the great enemy of these innumerable traditional techniques without which a country cannot live. Town culture will teach country children little more than how to read the serials in the newspapers. What do educationalists know about the countryman's self- and ancestor-taught education? The town talks about redeeming our agriculture: its way of doing it is by urbanising the country, drawing its labourers into the factories and depriving its children of their due training in husbandry and interest in country pursuits.

I believe that the only way of saving the countryside (and incidentally increasing its food-production) is by doing precisely the opposite of what in dozens of different directions we are about at present. There are, to my mind, only two ways of so doing – by a drastic decentralisation and by recreating a contented peasantry in small holdings, either paying rent to landlords or owning their own land. Large estates hang like millstones round their owners' necks; these encumbrances are sold off to building jackals, not, for their own and England's salvation, subdivided into peasant holdings. "The only way," wrote a man of wisdom and understanding – Lord Ernle – "of restoring reality, purpose and meaning to village life" is "democratic ownership." It must be that, he said, or no ownership at all except by the State. To one thing any person of common sense would agree without demur. The only method of keeping the countryside in being is not by urbanising but by countrifying it.

A Countryman's Journal (1939:150-4)

Land Conscience

I HAVE reached the end of my long and circuitous journey across some very difficult country. It is right, therefore, that I should here look back from a high place and take a bird's eye view of the winding road I have jogged and its many milestones I have passed. I began upon a personal description of the house that I built and the garden that I made on the borders of Oxfordshire. In so doing, I became more and more conscious of the fact that I was a migrant and an alien. The only prescriptive right I had to my land and my view was that I paid for the one for the sake of the other, and that, too, out of borrowed money. I had no inherent and local claim upon this land whatever. I was faced, therefore, by a heavy responsibility for an appropriation based upon the most trivial of all grounds – a financial transaction. My only real justification was the development of a kind of land-conscience, and I gradually came to see that this carried me a great deal further than originally I had any wish or idea of going.

First, there was an inward compulsion upon me not to violate the sweep of hill and plain that looked critically on at what I was doing, and would judge me according to whether my intrusion was persuasive or coercive. But, whatever I did and however diplomatic I was with the land to which I had so absurdly slender a right, I was still the alien cutting in upon an immemorial tradition of co-partnership between man and nature, who between them made England, the real, the beloved England. I was worse, I was a unit in a horde of aliens who were descending upon the English country as the plagues descended upon Egypt. I was an atom in a movement, a drop of water in a current which was tidally flowing out of the towns into the country. I was oppressed with the burden of the townsman and the just charge against him that he had starved the country when England began to be industrialised and now was choking it by the return to it. That was caused by many factors. But the most spiritually important one was his disillusionment with the whole fabric of a mechanised urban life and the bitter fruits of that progress that was the idol of our fathers. He was coming back to his ancient mother.

But, alas, she was not as she once had been. The village community whose origin is far older than mediaeval times, a mere yesterday, had received something like its death-blow from the process of starvation – the cream of it had all been skimmed off. The invaders on their part, all but a fraction of them, had no real bond with the country at all. Consciously or unconsciously they revolted from the towns and the Machine Age they represented but without belonging to the country any more. They had lost the key to the ancient, the enduring rhythm of intercourse between nature and man that poet and ploughman alike possess. To perform the semi-ritual of that intercourse there were left only the dwindling society of the labourers and a handful of public-spirited landowners, with here and there (very widely apart) a farmer or two – for the modern farmer is becoming a business man, and a bad one at that. There lay the decaying and more and more deserted countryside, and to its Canaan were crowding the new people, the Children of Israel of to-day. The silver cord was loosed and the golden bowl was broken, and for the first time in the history, and the prehistory, of England. We were wanderers through the wilderness and where was the sign to guide us, the pillar of cloud by day and the pillar of fire by night?

Through the Wilderness (1935:278-9)

PART THREE

1940-52

Progress – 1

THE modern theory of Progress is the archetypal example of abstracted thinking. Its fundamental deduction that human betterment is automatic and universal is clearly related to Puritan and economic determinism; it arbitrarily abstracts present time out of its historical context, regarding the past of human experience as "obsolete", and its standard of measurement is purely quantitative. The theory of Progress as the modern world interprets it would have been impossible in an age of concrete and realistic thinking. The extraordinary development of science equally reflects the dominance of the abstract outlook. Its principle is uniformity – the framing of general laws abstracted from the phenomena of nature; its method is the way of separation – the part fragmented from the whole. A good example of this scientific habit of mind occurs in the modern use of artificial manures. The problems of fertility are studied not in the field but in the laboratory, and chemical analysis of the constituents of the soil is presumed to solve the interrelations of living elements. But the same idea permeates all scientific enquiry. . . . It is no exaggeration to say that science in its "progress" has become more and more abstracted from the study of nature as it presents itself to our experience.

Wherever I turned in the kingdom of the mind, I found the same tendency omnipresent. The *laissez-faire* doctrine of the law of supply and demand which obsessed the 19th century is as abstract as it well could be. Money, which is simply the hyphen between production and consumption, is no longer the medium of exchange. It has become a science in itself, indeed a mystery cult, guarded from the vulgar scrutiny by an inner priesthood, a highly complex technology and an abstruse terminology. Mass-production, herd-mentality and standarisation of clothes, food, buildings, employment, education and way of life, all betray that quantitative assessment which is the essence of the abstract. Centralisation and bureaucracy are allied manifestations: in Germany and Russia, the State is abstracted into an absolute power which *per se* has no existence in reality. Mechanisation whether of things in factories or of thoughts in mass-suggestion is an application of that law of uniformity which guides the concepts of science. Nothing is more significant in our era than the inferiority in status of the primary producer to the dealer and middleman, and this distortion of values represents a departure from the nature of reality into abstraction. The abstract has no base; it is removed out of the context which in all work is creation, in all living is nature, and in all experience is life. . . . All modern thought in its over-cerebration reveals the same process of separating the part from the whole, the same sign of isolation from experience . . . Upon our age has been laid the curse of the abstract and, because it represents an escape from reality, it is manifestly breaking down. Too much abstractifying leads to madness, and has not modern Europe gone mad? In order, therefore, to save our civilization from total disaster or, if it be not worth saving, to save the human entity from modern civilization, there have to be alternatives to this tyranny of the abstract which imprisons us. As I see it, the return to realism, the rediscovery of concrete experience may be reduced to three primary elements – the Christian faith, individual responsibility and the land. The whole point of the Christian religion seems to me to be that it is the only historical one. . . . It also covers individual responsibility which, because it works from the bottom upwards, expresses the contrary principle to the abstract idea which works from the top downwards.

The third alternative is the land. It is impossible to visualise anything more concrete than the land; it is *the* context of our earthly life and the plinth or foundation of all civilised life. It is non-derivative, inalienable, and the source of human experience. Nevertheless, so fettered is modern man in the bonds of the abstract that he is attempting to apply to the land the quantitative principles that govern his urban life. The first or nearly the first English step in this direction as the Enclosures of the mid-18th century which coincided with the victory of the abstract over thought. The immediate consequences of the Enclosures were the dispossession of the peasantry, the degradation of the wage-labourer into pauperism and a large-scale capitalist tenant farming which regarded the land from the point of view not of livelihood but investment. The middle consequences were the application of machinery to the land, the growth of urbanism, the depopulation of the countryside and the treatment of agriculture as a competitive industry subject to the fluctuations of international trade. The ultimate consequences have been the insolvency of the farmer, the impoverishment of the landlord, the reversion of the land into second-rate pasture and scrub and, latterly, the attempt to mechanise it as a production factory and run it as a business "rationalised" like other businesses by science. The attitude to the land, that is to say, has proceeded step by step with the spreading and intensifying of abstract thought in urban life. The early and middle consequences have been failures; the final ones will follow suit for the simple reason that the earth is the one uncompromising datum in our mortal life that will not give in to abstraction . . . How remote seems mediaeval society from our own! Yet we are not so far away from it as appears. No society turns its back upon itself in order to become what it once was, however widespread the demand for sackcloth and ashes. What it does is to rediscover in a new way the eternal principles upon which our temporal life is based. The world has left the true values, the concrete realities away back in the "dead" past and, therefore, it will have "to put the clock back" in order to go on living, just as England has already put the clock back in order to go on eating. But if we find these realities once more, we shall not go back to what we were. Experience by its very nature cannot recapitulate; but it can and does and must restore, replenish, refill and reanimate itself out of those sources of life which the past made better use of than we do. What it does is to act as some organisms do, namely to upbuild from the simpler to the more complex and to break down from the excessively complex to simpler conditions. The complexity of our society is too heavy a burden for it to bear. Mediaeval society was intricate enough but not merely from without; its spiritual life was sufficiently developed to balance against its own errors, discords, cruelties and superstitions. Ours, alas, is so hollow as almost to form a vacuum. Consequently, its movement, abrupt or gradual, will be katobolic and towards simplification. In that sense, its direction will be from the modern towards the mediaeval, unless the disintegration is more radical. It will move from the bigger towards the smaller, from parasitism towards self-support, from centralization towards federated groups, from urban towards rural, from the factory towards the workshop, from mechanisation towards the co-operation of hand and eye with brain. At least, it will if it is to go on living. These excellent things are all in essence mediaeval.

Remembrance (1942:114-18)

Progress – 2

WHEN I talked chemicals and tractors to my father, said Hosking rather ruefully as we drove among the Downs, I was called unpractical; now that I talk against their abuse, I am still called unpractical. Before the first phase of the war period, he had been urging some compliance with "progress" upon his father who had five sons to place out in the world. Even so, it took the old man some years to be moved, not from what the modern calls a stubborn obscurantism but his intuitive sense of what he believed to be right. It was his son, not he, who took the line of least resistance. Being young, Hosking had caught the infection of a pace-making system entirely hostile to a self-supporting farm-economy. The opportunities in agriculture were becoming fewer and fewer, as society became more and more urban, and the skilled workers were flocking into the towns where they or their sons were to become either unskilled at the machine or workless on the dole. The first world war completed the parting of the ways, and, being one of the younger sons, he enlisted. He was absent till the pseudo-peace released him. Accepting the odds against his father's type of farming and suspecting that agriculture would be betrayed as soon as the cheap-food ships could get under weigh, he joined the business of his uncle, a miller-farmer, in the region of Okehampton.

It was on the uplands looking down on Mells, near which we had been to see another rick, that he told me the outlines of the second phase in his working life. Mells has poignant memories for me from my own earlier life. I looked down into that enchanted combe, so rich in trees, in the dove-coloured grace of the Horner mansion and the Somerset tower beside it. The secluded cottages were scattered like flowers in grass among its leafy slopes, and by the waterfall behind its merry stream where the dipper used to nest. Mells seemed to be an emblem of what Hosking, like our whole nation, had left behind.

On his father's side he came of a line of landowner-farmers, seamen and parsons. It was from his mother's side that the milling tradition was inherited. Her father had plied his own steamboat with its brass funnel on the same Cornish river in whose crook stood the parental farm. It carried the wheats which were loaded from the boat for the mill on horse-waggons and in those spacious days there were 17 of them. This countryman too lived beyond the three-score years and ten, and rode his horse up to a few days before his death. His son, the brother of Hosking's mother, maintained the family vigour and independence as well as the milling business. He had three mills, was four times mayor of his town and had insisted on addressing a hostile mob from his steam-waggon.

But in the middle of the nineteen twenties, both the Devonshire mill and the two family country mills in Cornwall, were bought out by the milling combine, then in the thick of its campaign for putting the country stone-grinding mills out of action and so striking a mortal blow at the heart of the localised rural community. It was abetted by Government who passed the paralyzing law to veto the local farmer from selling bread-corn to the local miller. It was not because these local mills were "uneconomic," that they were struck down. The costs were as low as the combine's. But the stone-mills were smothered by the underselling of surpluses and the erection of temporary plant in the district. Long before the twenties, his uncle had migrated into Devonshire. There the nephew after his war-service joined him in the milling of flour and grist, in the merchanting of seeds, wool and fertilizer and in the farming of some 300 acres on the edge of Dartmoor.

So short a step, so minute a change – from Cornwall to Devon, from father to uncle, from a farm that was just a farm to a farm with a mill. Actually, it was a giant stride from world to world, the stride of one century that our nation took after the Industrial Revolution from a rural civilisation into industrialism. For Hosking passed at one bound from self-subsistent production to mercantilism, from growing and conserving things to buying and selling them. The wrench was self-confessed in his telling me that he used to drive 65 miles from Okehampton to the Cornish farm but stopped four miles away from it. He knew that, if he had added this fraction to his mileage, he would not have been able to return. It was not home-sickness after four years' absence at the war; it was the pull of the land from which he had been uprooted. He put the meaning of the change himself in saying that the people who walk on pavements have lost the sense of heaven and earth. Of earth *and* heaven.

The economy of his father's farm had depended like the self-renewal of the animal organism upon its own powers and processes of self-maintenance; now he found himself buying wool from farmers in Devon and Cornwall, seeds from London, from Timbuctoo, from anywhere. And selling them again, trafficking in them. The buying and selling in Cornwall had been incidental to the farm, but now it was the farm that had become incidental to the exchange from without of farm-products. So with wool. How little, he said, does the wearer of wool realise how much the weather, the soil and the nutrition of the animal affects the quality of what he is wearing! Seed, as he said himself, now represented so many lbs. at such and such a price; he was no longer seeing it from the cradle, watching its potentialities and assessing its quality. It was in his blood to note what the seed could make of itself and what he could do to induce it to make the best of itself; it had become his profession to make what he could out of it. The sense of what seeds were and could be in themselves is stronger in him than in any man I have known, and the puzzled way he spoke of having looked at them from a totally different point of view was even more revealing than his tale.

Then there was the mill. This was a steel roller-mill, not the stone-grinding one his uncle had been forced to let go. The difference here was as great as that between growing seeds and only profiting by them. The stone-miller of course made his profit, but his primary purpose was to grind grain into flour. The primary purpose of the roller-miller is reversed; his rollers extract the wheat-germ because it is profitable to do so. Hosking was in no doubt at all about this. Long before the days of the breakfast-food firms buying the wheat-germ, he took it home with him for his porridge and his biscuits. To-day on his table he has the best whole-grain bread I have ever tasted, baked by his wife. The good miller, he said, milled the whole grain palatably, hygienically and digestibly from fresh wheat locally grown. But it would not keep (good things to eat seldom do). The modernised millers imported the dry Canadian wheats which will store and travel, standardised the moisture content and extracted the offals with the germ to be resold. Big business sprang out of lifeless bread. Because of their high moisture content, our home-grown wheats, superior in quality to any wheat in the world, cannot be stored for a prolonged period.

Both uncle and nephew were perfectly well aware that the whole-grain flour of the home-wheat had a dietetic value to which the imported wheats could not be compared. But by the irony of progress they found themselves buying wheats from all over the world at the expense not only of the old family business of watermilling, selling the flour locally and grinding it whole but of that very self-sufficiency of which the parental farm in Cornwall had been a shining example. At the expense too of nutrition. How important is this? We do not yet know. But we do know that when unpolished rice was stripped of its skin, beri-beri swept not only many a native population but our troops at Gallipoli. As soon as the whole rice-grain was issued to the troops, they recovered. The germ of the wheat lies close to the skin of the berry: both rice and wheat are cereals and the process of removing the life-force is much the same in both. The life-cell of any cereal is as the hearth to the house.

The Wisdom of the Fields (1945:214-17)

William Cobbett

IF Cobbett be cut to the measure of his own age, his life was a total failure. No sooner was "the most powerful political writer of the day," as Hazlitt called him, buried with his fathers in the churchyard of Farnham, than he was forgotten. Even his own people forgot him, as the countrymen sucked into the factories of the industrial "wens" were in one generation to forget the fields that had been their fathers' for a thousand years. Were this failure and this forgetfulness purely the consequence of defects in his character and education which were bitterly resented and ultimately brought him into disrepute?

On the modern reader Cobbett leaves an impression of a wild dispersion of energy. His lack of self-discipline creates a sense of waste and frustration, and he too often confounded his onslaughts on abuses with personal abuse of individuals. He was not the man to let I dare not wait upon I would, and never paused to ask whether it would be wiser not. He was just as self-confident about things he failed to understand as about those he knew better than any man in England. Often he hit right and left in a kind of frenzy, and charged as blindly as a rhinoceros. Among his host of enemies the worst was not infrequently himself. Though he possessed a style like a steel blade, he often used it like a club. The wind that filled his sails was partly that of a Titanic vanity. But it was purely personal and has an element of naiveté; his conceit had nothing in it of calculating egoism. He was a man of wrath but no Stiggins.

Often he was needlessly provocative. He rode round and round Dundas at Highclere merely to annoy a man whose politics he detested, and with his son went whip-cracking and hullabalooing to torment the Botley parson. When he sensed, and sensed rightly, that there was defilement in the air, he could not contain himself. He roared aloud and plunged without first probing into the details of the scandal. He was always the intuitive man, and a fatal impetuosity translated his detections into instant action. His attack on Peel in the House of Commons is the classic example of his rightness in smelling out corruption and complete wrongness in the details and direction of his attack. Peel had not manipulated the currency in order, as Cobbett said, to increase the value of his ministerial salary. But that manipulation for which Peel's Bill was largely responsible had fattened the new rich of whom Peel, for all his personal integrity, was a supreme spokesman. It is easy to see not only why Creevey called Cobbett "a foul-mouthed and malignant fellow" but how apt he was to arm his adversaries with both hands. Their weapons were his own mistakes.

But all this was not the ultimate cause for the hatreds his hates aroused. He was feared as well as derided and scorned. Authority did not heed his warnings and dark prophecies, but it trembled lest he should raise up the dispossessed against it. It prosecuted him in those fears and was discomfited. And if the Reform Bill had given votes to a determining majority of the English people, it is likely that some at least of Cobbett's aims would have been achieved. For the machine-workers in the new towns were still, when the Bill was passed, conscious of their peasant stock.

He was feared, too, because he told the truth about a number of things which almost everybody in public life, then and for generations to come, wanted to conceal not only from the public but from themselves. Cobbett himself cannot be understood unless the nature of those truths be realised. It is clear that he did not foresee the years immediately ahead of him. When he was riding up and down English country in the eighteen-twenties, reporting for his *Political Register* and haranguing a labourer met upon the road or a crowd gathered to hear him, he was usually a prophet of the immediate wrath to come. The wrath did not come. On the contrary, it appeared as though events had falsified his predictions. The intoxicating if most unequally distributed prosperity of the mid-Victorian age lay immediately ahead. The mood that heralded it was complacent and optimistic, and so a fertile soil for the doctrine of automatic and inevitable progress, unknown a hundred years before. Cobbett's truths were timeless ones because they were founded upon first principles. But time must take time to bring in their revenges. Their incidence fell not upon his contemporaries nor their successors but ourselves. Thus, quite apart from the personal antipathy he aroused among his opponents, the fear of him was mixed with the contempt felt for him who cries wolf, wolf, when no wolf comes. The superficial evidence was all against him, and *The Times* in a scornful obituary notice dismissed him as an episode.

Disraeli said in *Coningsby,* "the spirit of the age is the very thing that a great man changes." Cobbett was too great a man to be carried away by it, to drift with it, to bow to it as not to be avoided. He was not great enough to turn it. But the fact that he withstood it is a sign not of his weakness but of his strength. And the forces arrayed against him were gigantic. Would a man of greater wisdom, of a more formal education and of a superior strategy have prevailed against them? He stood against them like a bull stuck with darts in the ring, and he stood alone.

For his was the world of the new industrialism which Blake had foreshadowed in his "dark Satanic mills," and Cobbett's ex-commoners were being rapidly absorbed into it. It was a world that was beginning to think in terms of the inorganic and controlling the forces of physical nature. But when Cobbett thought of nature, he was thinking of the fields he knew, Little Foxhanger, the Seven Acres, Haw Croft, Priest Croft, Barley Close, Grunt Drove Meadow, plots of land that demanded individual treatment and had been named by his own people who for centuries had had a responsible stake in them. It was a world about to set forth upon "the conquest of nature," and Cobbett's idea of man's function was for him to co-operate with nature in the management of living growth. It was above all a period of expansion, propelled by the application of cheap power and the progress of mechanical inventions. The way of life that Cobbett taught was the very reverse. His view of this extensive movement is given in *Cottage Economy,* which is a treatise of intensive husbandry. This expansion, the opposite of conservation, was accompanied as time went on by social disintegration. Cobbett was an apostle of holding the fabric of things, both in society and among individuals, together.

The new machine-power was not in itself the cause of a

revolutionary attitude to old values and stabilities; it was a terribly effective instrument of it. The real cause was the dominance of a new state of mind which used the machine to accomplish its ends, and, as its powers grew, became more and more impatient of traditionally religious and ethical checks upon economic expansion. Those ends were the pursuit of wealth not for any specific social purpose but as a standard of value in itself. Cobbett's perception of this is at the root of his constant and often bewildering diatribes against finance. Thus ends were beginning to be lost in means and just the same process of confusion was happening to the machine itself. Progress began to be identified with technical advances and from this fusion arose the theory of a progress that was inevitable: that is to say, progress was in itself regarded as a machine, and so became separated from growth, which is an organic process. The effect of the machine went much further than throwing the skilled handloom weavers in the towns out of work and breaking up the cottage crafts that Cobbett loved by the migration of their rural workers to the new industrial towns. In opposing this drift and trying to reintroduce certain of the crafts, Cobbett stood like a barrier reef against a whole sea of change. But how great a change he hardly knew. The mechanisation of work went further than the creation of such squalor, poverty and slum-conditions as Disraeli described in *Sybil.*

On the one hand, it undermined the institution of the small property that, with the economic freedom it gave, was the corner-stone of Cobbett's faith. On the other, for the skilled labour of the craftsman and artisan on whose behalf Cobbett fought all his life, factory work began to substitute the unskilled labour of the machine-minder. The early reformers very properly welcomed the machine as a means to eliminating drudgery; what it actually did was to eliminate skill and pride in work and so create a new drudgery of its own, the drudgery of mechanising not only work but men. This alone explains the wonderful reception Cobbett received from the industrial workers. He did not himself think at all clearly about the advent of machinery. But the machine-minders of the North recognised in him the defender of a way of life that their fathers and even they themselves had enjoyed and contrasted bitterly with their present servitude.

Because the end of life was becoming the accumulation of riches, finance too had got out of hand and from a means became an end. It had made enclosure a profitable investment and was sweeping the land-workers off the land. It was beginning to ruin agriculture altogether since, when money was being used to breed money, it was being diverted from irrigating the growing of crops. Cobbett saw clearly that prices, rising year by year during the French wars, had upset the whole agricultural equilibrium and made land and agriculture "objects of speculation." Landlords refused to grant long leases because the value of money was bound to fall, while the labourers got nothing out of the artificial prosperity of the war years. Small farms were being absorbed into large. Just as in the nineteen-twenties prices fell after the war and deflation replaced inflation, the full weight of the depression falling on the farmers and labourers. The farmers of his day were living on borrowed money just like those of our century. And Cobbett did realise, as few enough modern farmers do, the inveterate hostility of finance and industrialism to agriculture. So, though not understanding the detailed workings of currency, he was sound in his demand for a steady one.

But the real point of his criticism of his age is that he saw the immediate issue as shaking the very pillars of society. It was because he perceived this stalking terror that a very *daemon* took possession of him. He prophesied that this financial "THING" would destroy the stable agriculture of a thousand years and in the fullness of time it did destroy it. In the poverty of the labourer he called his "chopsticks," he foresaw that the debt system would become a millstone round the neck of the nation and so it has become. He dreaded that, as taxation increased to feed an insatiable Debt, so security in and responsibility for property would result in loss of property for the many and much too much of it for the few. History has justified him. As champion of the "small man," he was vehement in denouncing monopoly. The growth of monopoly is a much more pressing issue in our own day than it was in his. But he showed his greatest historical insight in his crusade on behalf of a self-supporting nation resting firmly upon its biological roots. These agricultural foundations, he declared, were the only bulwark against the misuse and so the corruption of money. When a secondary economics overbore the primary vocation of agriculture, values went to the wall. While Cobbett was living, England was beginning to correlate her new position as "the workshop of the world" with the bulked imports of cheap food from abroad. He was thus considered as a stupid and sentimental impediment in the path of progress. But with us he emerges as more modern than the moderns.

There are two reasons why this is so. When Cobbett is considered as a countryman, his faults and failures sink into comparative insignificance. Secondly, the word "foundations" sums up the whole of him. He stood for foundations not in any particular period but in all periods and, though he was the most English of Englishmen, not only for one nation but for all nations. His modernism becomes a shining light because the modern age has threatened those foundations as never before in our history.

Cobbett has often been appraised as the guardian of the "commoners of England," dispossessed by the Enclosures. He has often been admired as the passionate lover of rural England, the England of the agricultural South that his piercing eye took in from the saddle. But he has been rarely if at all approached as reflecting and representing the countryman's point of view, whether labourer, farmer or squire. Though often in conflict and very differently placed in station and conditions of living, all three, so far as they fulfil their service to the land, do throughout their history share a common point of view. In his simple forthright English Cobbett spoke it. In fact, he embodied it. He translated it into his politics and economics, into all his ideas about man's social relations. But because the cottager was the foundation of the whole rural structure, as that structure is the foundation of the national life, it was to the cottager's well-being that he dedicated his arduous life. This perception, for instance, is the key to Cobbett's insistence upon the primary value and virtue of the family. His ruralism can hardly be called a philosophy of life because so much of it was instinctive. But it was certainly a way of life which with many vicissitudes had lasted ever since there was an England. The rural attitude to life, in which the whole English tradition is rooted, sprang to articulate life in Cobbett at the very moment when it was beginning to crumble and give way to alien forces. And it is through him that we not only understand what it was in his time but has always been.

He therefore truly interprets the soul of the countryman. The significance of all that Cobbett wrote is that it is countryman's literature. It is liberally sprinkled, for instance, with tale-telling, and telling tales has been the immemorial social habit of the countryman. It abounds in the colloquialisms and familiarities of country speech. It is concrete, realistic and based on observations of actual things. It is full of memories, and so is life on the land. It loves parables and illustrations as did the country Carpenter of Galilee. It makes plain statements as countrymen do and is as fresh as the fields after rain, even though Cobbett keeps on saying the same thing over and over again. To read Cobbett is to hear him, for his "literature" is that of the spoken voice rather than the pen, and this oral language is characteristically rural. It is an earnest and moral tongue, not to say didactic, and the countryside has always been a stronghold of the traditional English morality. And Cobbett's style is beefy and brawny as befits the style of a man who glorified the farmhouse meal and home-brew. Chesterton has said that there always are two songs in the Englishman's ears: "Home Sweet Home" and "Far Away and Long Ago." Cobbett's life fulfilled them both.

To discover the essence of Cobbett's ruralism we have to accompany him as he rides over English earth, casting on all sides the sharpest pair of eyes in England. It is by what he saw that we gather what he thought, what he taught and what he believed. Much of the *Rural Rides,* written between 1821 and 1832, is descriptive of the countryside from the saddle. But Cobbett's eyes never saw the landscape alone as a picture; they saw everything that grew and walked upon it, and not only these but the politics and economics they signified. In the valley of the Avon, for instance, he counted 7000 sheep in a parish of 3500 acres. The population was 530 and he calculated that it produced enough bread, bacon, mutton, wool, game and dairy foods to maintain a community of 2500. This is a good example of how he used a particular observation to illustrate an economic issue of the widest significance. In the same way, his eyes never separated what was useful from what was beautiful, and so all traditional countrymen have seen them. They are bound to be so seen because nobody is in a better position to realise that English country is man-made or man-husbanded than the countryman. The English countryside from prehistoric times has been man at one end and soil at the other, and this organic sense was the inspiration of *Rural Rides.*

A tree was not beautiful to Cobbett if it was covered with ivy or moss, if its growth was crooked and its heart unsound. He hated firs and spruces ("beggarly stuff") in inverse ratio to his partiality for the oak and the ash. The picturesque attitude to natural beauty would find this meaningless. But firs and spruces are not a native part of England's flesh and bones like Kipling's oak and ash and thorn. Long before Cobbett and Kipling, Wordsworth and Cowper, they had been the very stuff of livelihood, trade and beauty to the long generations of nameless craftsmen. That was how Cobbett saw them, both rising from the earth and lying in the workshop. The abstract view of beauty as a thing in itself, plucked out of its context in life or truth, in use or virtue, meant as little to him as his view of landscape to the votaries of the merely pictorial. It would be hardly possible to pick out of the whole of Cobbett more than half a dozen descriptive passages in which he looks at the country apart from its husbandry. He has, for instance, many references to the birds of the countryside. But the kind that figures far more frequently than any other is the rook, the bird that follows the plough.

His way of seeing beauty and use as two halves of one whole appears just as conspicuously in his attitude to the human no less than to the natural scene. At Great Bedwyn, he saw a bevy of comely girl-reapers in rags, and rode on to curse the sight,not to praise it. Cobbett had a sharp eye for a pretty girl, but beauty for him was a mockery with misery behind it. It had to be heart-deep; he had no eyes for it when only skin-deep. "Handsome is as handsome does" was his aesthetics. And if he saw with his naked eye a fine and fertile valley, he peopled it in his mind's eye with "property-holders," squires, yeomen, farmers and small-holding labourers, redeemed from the National Debt. A skimpy land, Bagshot Heath for instance, had no appeal for him at all, while he was always looking for, though seldom finding, a rich soil supporting a prosperous people.

Particularly in the cottage gardens the various aspects of his vision became harmoniously fused. "These neatly kept and productive little gardens round the labourers' houses" are "an honour to England" and to be seen nowhere else in the world. "We have to look at them to know what sort of people English labourers are." And English country itself was to him the cottage garden of the world. . . . For wild natural beauty, untenanted by man, he had little or no appreciation. His views thus diverged sharply from those of the contemporary "Romantic Revival" which had little thought for a domesticated Nature with Man as her partner. A smiling land and a smiling people living on and by it, this was his earthly paradise.

The Wisdom of the Fields (1945:14-22)

Coke of Norfolk

TWO events this April end have put new heart into me and made it in key with the green outdoors. One of them is a fine example, taken from the past, of the compatibility between tradition and advancement in agriculture. When it is too late, it will be recognized that one of the major insanities of our time is the utterly artificial cleavage between them. If the unnatural hostility between them did not exist, we should not be spectators of the fantastic fact that the country teems with agricultural colleges, scientific theories about tillage and a weirder assortment of agricultural machines than even Swift could have imagined for his crazy Laputa – and that there are next to no men to do any tilling. All men, I presume, would grant to Coke of Norfolk a place of supreme eminence in the annals of husbandry and a faithful discipleship to Voltaire's "the best thing we have to do on earth is to cultivate it." Indeed, the story of what Coke did with his land at Holkham, a wilderness part sand and part bog, is one of the most brilliant and romantic stories in all history of what human goodwill is able to perform in constructive effort against adverse conditions and bitter opposition. How blest he was in dying before the Industrial Revolution had sown the dragon's teeth of our present impotence in benefiting and blindness in destroying our native land! He is invariably regarded as the great innovator – as, for instance, in introducing potatoes and the drill into Holkham – but nothing I have read about him (including a tediously lengthy biography) gives the smallest idea of what he really did. That was to restore in one region and under his own personal guidance the old traditional spirit of husbandry.

In 1805, Coke said in the House of Commons that "in Norfolk, where farming was carried to a great degree of nicety, he believed that there was no such thing known as the use of oxen in husbandry." Accordingly, he reintroduced them and harnessed his imported Devons, which he found superior to Shorthorns, to the plough. Arthur Young's *Autobiography* records that "in 1784, Coke worked twelve oxen in harness for carting and found them a very considerable saving in comparison with horses." In spite, too, of much opposition from the scientific folk of those days, he clung to the old local Norfolk plough and in ploughing matches with it defeated every competitor that challenged him. These are both instances of revival being a direct cause of improvement. He also perfected the rotation of crops and without it as an organized system (admittedly in a more primitive form) the old village community could not have existed. He used marl and clay for manures, quite discarded to-day, as the Celtic and Saxon farmers had done before him, and I may add that I have never met a really knowledgeable farmer who does not regard with suspicion the excessive use of artificial manures as one of the principal reasons for the impoverishment of our soil. Fifty years later (in 1847), "Talpa" (C. Wren Hoskyns) described in the "Agricultural Gazette," in which the "The Chronicles of a Clay Farm" first appeared, how he dug out the subsoil of marl and clay from his waterlogged and unproductive fields, spread it over them during the winter, harrowed them in March, and produced rich crops. His account of the scepticism and prejudice of the neighbours in face of this *innovation* exactly parallels the experiences of Coke.

He reclaimed land from the sea – just as the Cistercian monks reclaimed it from the Yorkshire wilderness. He recovered the waning industry of home-grown flax and hemp (finally destroyed together with a multitude of other crafts allied to crops by the Industrial Revolution) and founded the Thetford Wool Fair in more than memory of the great days of the fourteenth century. He fostered the alternating rhythms between work and play on the land, and restored some part of the older rituals of the countryside by making a great annual festival of sheep-shearing and enriching it with his liberality.

These are all quite definite examples not of introducing something totally new, as our infatuated notions prescribe, but of re-animating something that the commercialization of the land had rendered moribund or extinct. But what Coke of Norfolk did went deeper than in particularities. He acted generously and consistently upon his conviction that the interests of landlord and tenant were identical – and that was the basis of the manorial open-field system, however flagrantly the principle was sometimes abused. His whole life as a landlord was to practise towards his tenants the spirit embodied in the charter of the Peasants' Revolt in 1381, and his maxim, "Live and Let Live," was actually the same as theirs. And he earned the approbation of Cobbett – the arch-representative of the older England – who, after attacking him for his wealth, wrote in *Social England* (1818), after visiting the Holkham estate, "Everyone made use of the expressions towards him which affectionate children use towards their parents." He made of Holkham a local community in which none were indigent and none idle, and all in their several degrees made contributions, like the different parts of a Gothic building, to the *local* whole. Coke was a local prince and his great work was founded upon the Enclosures. Nevertheless, it is true to say that by his enlightenment and disinterestedness he was but improving upon the old stock of the local village community with its co-operative farming combined with individual ownership or tenancy. To us this story is largely unintelligible, because what is called "scientific agriculture," while increasing what is crassly called our "conquests" over nature, has paralyzed our powers of perception.

There is an uneasy suspicion abroad to-day that "progress" is "a tale Told by an idiot, full of sound and fury, Signifying nothing." Yet this is too pessimistic: progress does mean something, and what I take it to mean is a sudden longing, in the midst of confusion and tribulation, for the good things, the kind traditions, the old rhythms of life that the age has wantonly thrown away in pursuit of the power and the wealth that turn to dust as soon as in the hand. But what people say is – these things are fair and desirable and may even be our salvation, but there is no putting the clock back, there is no retracing our steps even if they lead to the edge of the cliff. They would acknowledge, for instance, that it is good sense, good morals and even a necessity enforced by pursuing these will-o'-the-wisps, that a country should feed itself without depending upon imported foodstuffs inferior in quality and nutrition and at the same time preventing its own countrymen from raising home-grown crops. If they knew any history, they would agree that the best means of securing this end would be to restore the land to its husbandmen and not to leave it either to revert into desert or in the hands of business men in charge of factories, of mechanics in charge of machines and men of science in charge of laboratories. Thereupon out tumble the old catchwords about the clock and the steps and the impossibility of returning to the past.

The Sweet of the Year (1939:108-12)

Living Soil

THERE is George the cow-man pitching a forkful of weedy hay (the sward is too poor for an aftermath) out into the field for the sheep and cattle. He knows the reason why they have no after-grass nor swedes nor mangolds nor kale nor anything good to eat except a wretched truss of hay. He knows, too, that the real reason does not appear in the newspapers nor over the wireless, and that it goes back to the times when England deserted her own country for the glittering prizes, the fat illusory gains and empty comfortable pickings that were to be got out of the race for wealth overseas. He cannot put it all into words, but the essential significance of what he is doing is not lost upon him.

Let me try to translate and interpret what is going on in his mind. First of all, he is thinking of a farm, the only one within a circuit of ten miles, where there are cattle and sheep, poultry and pigs and horses, all of them well fed. He knows why: it is an old-fashioned family farm with only a little machinery on it, partly in grass and partly arable and roots, and pursuing an out-of-date rotation. The farmer sits tight; he has been polite to the commercial travellers who kept on importuning him to buy their foreign cake and concentrates and artificials. He had no need to buy them because, in the face of every discouragement and adversity, he has gone on growing all that was necessary for the health of his land and stock himself. He has not stuffed his soil with expensive chemicals because the fertility of his farm is self-made, supplied by the beasts and the crops themselves. How different does his farm look than others in the neighbourhood, including the one on which George is working! For the past eighty years, and never more so than at present, we have treated the soil as dirt, as Henry VIII treated his wives – or, as on the land that George serves, we have degraded it from wife to drudge.

We have abandoned it to Suburbia on the one hand, faked and ginned it up with chemicals and wrenched the goodness out of it by mechanical contrivances, on the other. The peasantry, which traditionally understood it, we have driven off the land; we have regarded it as the raw material of a factory, while science, with its usual blindness to all but material factors, its analytic as opposed to the synthetic point of view, its confusion of half with whole truths, has occupied itself with nothing but devices to whip the last ounce of productivity out of it.

Since the soil is not dirt, but a living organic being, like man, dog, fly, it has reacted like an oppressed man, a starved and beaten dog, a fly with its wings torn off. It has been denatured, and between neglect and exploitation sinks into greater and greater infertility or, as we ought to put it, into apathy and sickness. Unhappily for us in our acutest need, a sick soil means sick plants, sick beasts and sick men. How many of the diseases produced by our civilization are the consequence of malnutrition?

Is there anything on earth more mysterious than a compost heap, piled in this farmer's yard? In the cycle of time, the rhythm of the seasons, by the transmutations of nature, by the secret operations of dynamic life, by the play of cosmical forces, that which was a plant has become the soil it grew from, and that which was dead has risen from the grave of itself to produce more life. That (not to mention abundance of skilled labour) is the cure for the malady of the soil, for the living dead to replenish the living earth, for decay and renewal to be a turn of the wheel of wholeness. The ancient Egyptians believed that the ear of barley was a god born of the womb of the earth – surely an idea nearer the truth than the way science, political economy and bureaucracy regard the soil.

The family farm, the antithesis of all the recommendations in the Astor-Rowntree Report, this archaic relic, the non-mechanized, self-supporting farm of our forefathers and balanced between arable, grass and livestock, this is the only farm which obeys the biological rhythms of nature, that can produce more food than the most highly organized, large-scale farm run by "joint-stock enterprise", and is virtually independent of the boards, the bureaucrats, the institutes, the financial deals with foreign imports, the very perils of the sea. At least it would be if it were let alone.

These are the realities that present themselves to George the cowman and so by a natural transition he thinks of their obverse. He feels in his bones that there is a cardinal falsity somewhere, the surrender to a vicious principle in a nation sacrificing its own sacred soil, the land of its fathers and makers, the bone of its existence, the mother of all its energies and the nurse of its character, for the parasitism of being fed at the end of a long tube. When war nibbles at that tube, like a shark tasting the diver's oxygen line, the weakness of such contempt for the motherland is clear to him. And in an obscure way, though without formulating general principles, he realizes that what the old husbandry stood for is being proved upon our bellies to be more stable, practical and fundamental than the whole intricate edifice of progress built up since the Industrial Revolution. We may go further and divine that some law, some power, some universal truth that science professes to have abolished, will not tolerate our violation of the essentials of life and is now brusquely recalling us to the verities. They are the earth under our feet and the sky over our heads, our daily bread and the spirit that is more than bread.

The Fall of the Year (1941:86-90)

The Mixed Farm

THREE issues, deeply involving both the past and the future, are so closely intertwined as to make a composite unity. They are the balanced or mixed farm, the health and fertility of the soil and intensive production. That all are characteristics of our traditional husbandry calls for no exposition. The first and second interact for the obvious reason that the mixed farm of arable, pasture, livestock and woodland is the only one which can restore to the soil what it extracts from it, and recognizes the basic wholeness of Nature by its own correlation of parts. Nature is the first of farmers; the mixed farm is the nearest to Nature.

It is a very curious fact that, though the specialized and mechanized farm, the modern version of the Roman *latifundia,* actually produced less yield of crops per acre than the mixed farm (large or small) and the smallholding, yet output and again output (the Research Institute, the government office and the business firm only think in terms of output), is put forward as *the reason* for supplanting the mixed farm and smallholding by "large-scale economic units." The figures prove it beyond controversy. In Sir Daniel Hall's *Reconstruction and the Land,* the spearhead of the "land as investment" attack upon our time-honoured husbandry, a table is given of the production per acre of wheat, barley, oats, potatoes and sugar beet between 1927 and 1936 in eleven countries. In every instance and by wide percentages, the list is headed by Denmark, Holland and Belgium, the type-examples of peasant and yeoman farming, and bottomed by Canada, Australia, the United States and the Argentine, the type-examples of large-scale mechanized farming for profit.

Peasant and yeoman farming is, of course, always intensive (when not crippled by finance and saddled with debt) partly by the stimulus of ownership and partly by that element of personal care and intimate knowledge which a farm run by machines and for profit alone necessarily lacks. It lacks something else as well. A fifty-acre holding properly cultivated and fed with humus has an internal surface equal to one of 150 or even 200 acres, so that throwing a number of smallholdings into one large one is actually wastage of soil. Thus, the relentless revolutionaries of "progress" are in flouting the tradition of the past mortgaging the welfare of the future, and from this conclusion there is no logical escape. Those very countries at the bottom of the list are now a byword throughout the world for the soil-exhaustion and soil-erosion of millions of their once fertile acres. In material terms alone and at a moment of supreme crisis in the relation of population to food-supply, the traditional husbandry of the country types I have been considering emerges as the one safe anchorage against a potential world-starvation. To say nothing at all of Sir Albert Howard's contention that a sick soil (viz over-driven or denuded of organic humus) is ultimately responsible for disease in plant and animal and a human population subnormal in health.

Since the assailants of the English tradition, nowadays so vocal and powerful, only refer to the question of values as one merely of "sentiment" and "nostalgia" and so presumably as out-of-date as private property in land, why should I refer to them in at all? Because the history of our peasant, our squire, our land, is pregnant with them, inseparable from them. The beauty of our landscapes and villages and little towns chimes with the inward beauty of those who have lived in and upon them, ranging the whole gamut from Shakespeare to the ploughman's song. A competitive world riven into chaos and going up in flames would seem to stand in need of that co-operation the peasant expressed not for emergency but in normal usage. A humanity perilously near the servile state and the automatism of the ant-heap has something to learn from the independence of the yeoman. A society in bondage to machines whether of the body or mind might look with something like longing upon the creative individuality of the craftsman and from the rapacity whether of countries or corporations turn with relief to his absorption in the worth of his work.

Is the individual leadership of the squire to be despised in a world of State absolutism and mental abstraction? Are the Christian values of Herbert, Barnes and the "povre Persoun" outmoded by the chaos of their denial? And is the quality of being exhibited by all these actors upon our stage not an alternative to the sense of mass which levels all things and persons to a row of numerals? Or, if we pass from the men to their work, material and spiritual craftsmen all, is not craftsmanship the true antidote to a humanity almost insane from something more than economic frustration? All the amenities of the most Utopian of Leisure States, all the ideologies from Communism to the Corporative State could never compensate the human being for the loss of individual craftsmanship in work. These things are the illusion of the urban mind.

I take it, then, that the English tradition of which our native husbandry is the base, being as it is true to the nature of the Englishman, is the concern of the future rather than a pleasant tale of the past. The fatal error of the modern theory of progress is that it interposes a destructive wedge between the past and the future, and at the same time regards itself as almost immune from criticism. A recent authoritative book has this sentence: "No sacrifice is too great to check the black plague (Nazi-ism) that would set us back to the 13th century." One can but wonder at the complacency of the modern illusion that considers as lower than its own barbarism the age that perfected the guild system and created the Early English style. The qualities I have been reviewing are always in date and only out-of-date at the expense of the human living-power and the national well-being. They are the conditions of a permanent satisfaction in the act of living; time may modify them to meet changes of circumstance but to destroy them is suicide. We cannot imitate them by "putting the clock back" because we have lost the knowledge of and the key to them. What we can do and what the times we live in will compel us to do, if we wish to continue as the English nation, is to recreate the English tradition in a form compatible with certain modern developments which are real and salutory, not fictitious and thoroughly harmful as many of them are. We do not want more bureaucracy, more regimentation, more commercialism, more bank-debt, more machinery, more competition, more complexity, more urbanism, more organization, more parasitism, more occasions for future wars, more and more imitation of the totalitarian or the communistic State, since all this is quite simply the way of death. The business of modernism is to fulfil the English tradition not to destroy it.

If we desire to have more life and have it more abundantly, we have to control, not to give full reign to these tendencies and to reassert our native humanity, our particular Englishness in spite of them. The example of the English tradition is before us who have become its prodigal son. Whether the inward rottenness of modern civilization as a whole has gone too far or whether the strain of suppressing the State-cum-military despotism of the openly predatory nations will exhaust us too much, cannot be foretold. But the effort must be made for our redemption as a nation.

The English Countryman (1942:133-5)

Norfolk

SLOLEY is a remote parish between Wroxham and the sea, and part of some of the flattest land in England. Ten feet above sea level there is a notable height. It was once well wooded, chiefly with oak, and these oaks were also hedgerow trees forming shady avenues on the surprisingly numerous by-roads. The roads form such a network that the visitor is constantly getting lost on them. But they are a great advantage to the farmers, and my friend's farm in the parish of 550 acres has roads round all the fields except for one or two outliers. Most of the land is in arable barley and the square or oblong fields are seldom less than twenty acres in extent, often more. Now that many of the woodlands and coppices and hedge-oaks are felled and the cottages, manors, barns and bartons barbarously decayed, it might appear from this survey that a region thus confined to few features is dull and monotonous. Moreover, the climate is harsh and winds from the North Sea sweep over the flats like a pack of wolves in full cry. Yet though I saw Sloley and its adjacent parishes in bitter weather and dominated by an east wind like showers of arrows, the impression it left upon me was warm and pleasing.

For one thing, the patterns of the oak-embroidered fields and the winter tracery of the trees themselves made a gallery of etchings and the land itself, in spite of neglect and exploitation, is still fertile. Nothing looks duller than an exhausted soil, however pictorial its environment. Much fruit of all kinds is also grown, and no country can be dull with orchards. And the place is remote enough to have been scarred but not branded with the intrusions of progress. It still looks itself, if that self be mutilated, and a place that looks itself is always interesting. Much, too, of the regional style survives for the same reason, if much dilapidated. The local materials are brick, cob and flint. A ruined smithy, now only bare walls, was built entirely of cob, and cob was used in the barns and other buildings above the wall-plate for securing the thatch at the eaves. The churches are mostly of flint, the secular houses of brick of a very good quality – sandy rather than rose-red in colouring. The manor houses, usually also of brick, often have stepped or rounded gables in the Dutch, or rather Flemish manner.

All these buildings reveal a sure sense of line and mass and, like Cotswold buildings, depend upon them rather than upon ornament which is restrained and sparse. Perhaps the most striking of the buildings are the barns, of which there is a considerable number. They are of plain sand-red brick, without porches, extremely simple in structure and always of just the right length. Always, too, they are reed-thatched, as the farmhouses and bartons usually are. Norfolk reed-thatch is world-famous, and here with the Broads within easy reach for the reeds it excels in its own place and element. My friend's barn which he is re-thatching has had the original reed on it for 200 years. It stands next to the manor-house, once the home-farm, also reed-thatched but with rounded gables beside the sharp, long, thin, triangular gables of the barn, surmounted by a blunt finial. Likeness and distinction exactly marked the fact that one was a barn and the other a farmhouse . . .

By a stroke of dramatic irony, there is to be seen at North Walsham one of the last survivals of the pre-industrial ages when the regional community and personal responsibility to the work and to God were a living reality, for churches now are hardly more than museums. This survival is the cluster of workshops belonging to and built by Farman, the reed-thatcher. He has had a remarkable career. The family may be called one of the most blue-blooded in Norfolk, Farmans having been reed-thatchers from father to son in an unbroken line since the twelfth century. The present Farman was prevented by the first world war from learning his father's trade. When he returned home to his father's house, he knew no more about reed-thatching than was in his blood by the inheritance of acquired aptitudes. His first job was to roof a house that was being built by an architect. The architect asked him how many square feet would be covered by the thatch. Young Farman had no idea, and the architect told him that without these measurements he could not have the job. Farman went home and worked out the sum until four o'clock in the morning, being five square feet out in his reckoning. To-day, he says that the architect did him the best turn of his life by compelling him at the very outset of his career to learn at once La Fontaine's maxim – *si quelque chose t'importe, ne la fais point par procureur.* A lifelong adherence to the principle of self-help has made Farman a successful and prosperous man . . .

The only machines I saw in Farman's group of workshops were an electric drill and a wire-cutter. All his manifold businesses are done by hand and in the old traditional way, there being no other. For what happened was that, as the regional crafts fell one by one into desuetude, Farman revived and re-established them by adding them on to his reed-thatching business. His workers ceased to be masters owning their own tools and workshops, but at least their trades were plucked out of oblivion. He has thus become a Jack-of-all-trades in most of the crafts that convert natural growths, including timber, a curious example of a combine which has *not* become so by amalgamating a number of separate and individual small businesses and dispossessing their owners. In our days, a creative and beneficent combine is the sharpest of paradoxes. Up to the last war, Farman was employing as many as fifty-five workers, a number that has now sunk to about thirty, since contemporary conditions, conscription and an office-bound bureaucracy among them, stultify the enterprise and organising capacity even of a master-craftsman of Farman's metal. A further irony is that in reed-thatching these men are seldom employed in their own county wherein the trade was supplied by nature with an inexhaustible raw material and was practised for centuries. The beautiful and elaborate, durable and decorative art of reed-thatching is pursued by Farman's craftsmen in Devon, in Cheshire, even as far away as Toronto, anywhere but in its own proper place.

In the pre-industrial culture, however, reed-thatching was the roof-mode of practically every type of building, except on the larger churches and manor houses and especially in the particular region where Farman lives where acre upon acre of reed is within easy reach. The roof-ridge is not of reed like the rest of the roof but of sedge and so is sharply differentiated from the high-pitched sides. The division is stressed by difference of plane; the edge being raised above the reed. This edge is often ornamented by projecting lozenges, oblongs or semi-circles. The reed itself excels by its velvety smoothness of texture and its deep warm colouring, a rich plum shading off into sepia, Vandyke brown and rust-red, entirely different from straw or rye thatch. There is no ornament at the eaves or round the dormers and the reeds are never cut, being pushed up into level lines by the thatcher's bat. The interiors of the barns used to be latticed with the reed between the rafters. The reed on an East Norfolk barn was made the more secure at the eaves by the butts being embedded in wet cob at the "tilt" or timber purlin running parallel with and about a foot above the wall-plate.

Since his own county now prefers blue slates from Wales, asbestos or corrugated iron for a roof instead of its native reed-thatch, Farman now does reed-thatching only as a by-product of his business. Consequently, when he took me along the lane outside his house with the workshops ranged on either side of it, I was seeing a kind of bazaar of the crafts. Here in a glorious confusion of diversity in materials, design and mode of workmanship were osier hurdles, gate-hurdles of split hazel, wattled hurdles, reed hurdles secured between three parallel poles, together with specimens of a dozen other crafts nearly allied to or at far remove from hurdle making. Farman makes even a fifth kind of hurdle, a riven-oak paling with the upright slats overlapping in the manner of weather-board. Here, too, were osier setts ready to be planted out in Farman's own osier beds, for he is one of the few surviving craftsmen who still owns his own osier beds. How my dead friend, Tom Hennell, would have delighted in this prodigality of natural resources and the multiple resource in their convertibility! I bought, for instance, an elegant little shopping-basket woven out of the Malacca cane left over from the cane-seat this Proteus among craftsmen also makes and, what is more remarkable, has trained his men to make. He will also execute orders for rush mats and the indestructible rush basket, though working in rush is almost extinct in England. Brooms, too, he makes and he has even built a workshop I saw in his yard entirely out of reeds for walls, door and roof. Add reed mats, garden chairs and tables and the lane might have been called, Arabian Nights fashion, the Street of the Artificers.

Such versatility – for Farman can make all that his men do under the personal supervision of himself and his son – may well be unique in this country, the graveyard of craftsmanship.

When it is combined with the specialised faculty of building up a business with so many ramifications, it will be seen what a prodigy Norfolk possesses in this citizen of North Walsham. Managing a collection of machines and machine-minders in a factory is child's play compared with the direction of a number of highly skilled crafts and the skilled handicraftsmen who work them. In a compound business like this, where the workers are necessarily individuals expressing themselves in their work and the master is himself a workman in personal contact with them and with a practical knowledge of all the varieties of work done, there is nothing to remind us of modern conditions in industry nor of the envenomed strife and ideologies that have sprung from them. Farman's business is far closer to the master builders' guilds which built the flint churches than it is to the modern impersonal factory.

Perhaps for this reason Farman remains obstinately Norfolk and provincial. He would not have me speak of "spars" or "spargads" for the thatching pins but "brotchets," and he was liberal with vernacular terms like "riving" for splitting and "flaking" for the special ornamentation of reed-thatch just above the eaves. He could identify twenty-seven different kinds of osier simply by moving their stems across his lips. And he talked Norfolk and looked Norfolk with the lantern face, shrewd twisted smile and loosely jointed limbs of Norfolk man. I ought to know because I am pure Norfolk myself. How happy he was to be given the job of re-thatching one of my friend's barns! The cost would be £300 – very cheap as thatching goes nowadays. But its cost was not what counted. What rejoiced him was the chance of rethatching a *Norfolk* barn . . .

My purpose in making a winter journey into these Norfolk byways was to see how a friend of mine was getting on in a modern version of the Labours of Hercules, reclaiming a farm of 550 acres from the devastation of progress. Nearly a year had passed since I had last seen the farm and then it had seemed that nothing short of a fortune would suffice for building up something solid and productive out of the ruins. My friend lacked that fortune and had in fact had to mortgage the farm to pay for his labour, stock, seed, equipment and repairs. He had the advantage of good land but that land had gone hungry for humus for thirty years. It had been regarded purely as a factory for cash-cropping and everything, soil, farm buildings, gates, fences and even labour (for the farm workers were so many instruments of production) had been sacrificed to that end. Modern economics had also starved the farm of the capital necessary to maintain it.

One of the inevitable consequences of efficiency farming is the elimination of the human element, in respect both of man-power and the human attitude to the cultivation of the land. It is concerned with cost accountancy in terms of machine-power and, so far as human labour enters into the calculation of profit, loss and time-saving, it is interpreted on the mechanical basis of machine drivers and unskilled casual gang-labour. A farm well manned with human labour is one of the worst heresies of our economic orthodoxy. My friend's purpose in taking up the burden of the farm was to restore the land and all that pertained to it as a self-regulating organism. He is, that is to say, a pioneer on behalf of traditional husbandry, since, whatever its shortcomings, traditional farming was always organic. Consequently, his first and main objective was to gather about him as many skilled men as the land would hold. At the end of the year that had elapsed since I had last seen the farm, he had sixteen men cultivating his mixed arable and pasture farm of 550 acres. This ratio of man-power to acreage is, of course, far in excess of that desiderated by the efficiency farming of "output per man."

But human labour considered merely as man-power, as, that is to say, an instrument of production, was no good to him at all. It was that kind of man-power that had left his farm in the desolation it was his aim to redeem. He wanted men who would co-operate with him in the work of rehabilitation. The only way to find them was first among the real old-fashioned provincial types and, secondly, among those rapidly increasing townsmen who have left the country by economic duress, but not in heart, or have discovered for themselves the deceitfulness of the industrial town. Since the latter class is usually unskilled like the vast majority of factory workers, most of the men he actually obtained were locals and herein lay the great advantage of my friend's farm being situated in an area of the country where towns are few and comparatively distant.

It was the custom of traditional husbandry to put the land first and self-interest second, and only among by-roads are the lingering traces of that loyalty to be found. But according to his ideas, the conception of co-operation would lack body and fail in inducement unless it took shape in a concrete form of partnership. The farm still being part of an estate, there was no lack of land or housing, however run to waste the one and dilapidated the other. Hence he was able to give his men cottages more or less rent-free and a pightle of land attached to each one. The men, too, had as much milk and other farm-produce as they needed. This rooting of the men to the land they served was certainly effective. One of the happiest of them had given up a job at £6 a week in London to come to my friend who gave him land and cottage free and a small wage. Later, when the work of regeneration had acquired momentum, he raised the wage of this man to the statutory minimum, and the man on his own initiative offered to pay for his milk and fuel.

Of course, workers on the land must have their roots in it: only a completely rootless society could ever have ignored such a truism, and then wondered why bribes and amenities were straws against the ceaseless drift from the land. Men so drift for the obvious reason that they have no anchorage in it. But when I saw my friend's men and talked with some of them, it seemed to me there was more to their feeling about the farm than that each of them had an individual stake in it. The "social contract" between master and man went deeper than a pooled self-interest in it, though that very naturally played its part. Why, for instance, did the foreman and the teamsman turn out in their Sunday best on the morning for breaking in the colts? Not only because they had their cottages cheaply and pieces of land of their own. Not only because they had a master who considered them, was anxious for their welfare and said to them, in the manner of Sir Thomas Overbury's yeoman, not "go to field" but "Let us go". Not only because their master lived with them and worked with them, himself tramping the land, learning the farm jobs, helping to build up the walls, load the carts, pitch the hay and repair the buildings. The bailiff, for instance, of a yeoman stock bedded deep in Norfolk soil by the successive strata of generations, had become a personal friend of the owner and took his meals with him at the home-farm. All these deeds of a valid partnership exerted a personal influence over the men of incalculable potency. But still there was something left over. I can only call it a traditional sense of loyalty to and responsibility for the land itself. They were pleased at being liberated from what they called "tear and rip" farming. Such ceremonial occasions as dressing up for the colt-breaking are, of course, a rural tradition that arose out of the loyalty of the local community to its own land. Here, the tradition was reincarnated out of conditions favourable to it, a very different thing from a mere time-lag. This was the source of the men's willingness, diligence and air of contentment.

I invariably saw on this farm what is not to be seen throughout the length and breadth of the countryside except now and again and in out-of-the-way places. I saw what is utterly unknown in any modern factory. That was men smiling at their work, an extraordinary experience. They did not only smile when the boss accosted them or when talking among themselves. They smiled while they were working and this was a revolution very different from the current trend of revolutions. If men could smile at their work once again as they did when they were responsible craftsmen, there would be no more revolutions, for the men who make them are those that frown or yawn. No wonder these men smiled. They were bringing order and beauty out of the ugly litter of efficiency farming. If a man on this farm was blamed, it was for going too fast, not too slow, for the speed of industrial farming is the deadly foe of thoroughness. Now men were working for the land as of old they had worked to build the flint churches that are now falling stone by stone into ruin.

With such a team of men at his back my friend had won half his battle of reclamation before it had well begun. Not many miles away, his great predecessor, Coke of Norfolk, had worked in his smock alongside his tenants to make gold out of sand. Whether or no Coke originated the Norfolk axiom of "muck is the mother of money," he certainly acted upon it, and so transformed his sandy barrens into rich cornfields and fattening pastures. My friend proposes not only to restore the old Norfolk saying derived from Flanders – "No keep, no stock; no stock no muck; no muck no crops," as the gospel of his farm but to go one better even than Coke by winning his men over to compost. For his object is to double the fertility of his land, and the difference between himself and the industrial farmer is that one makes the land serve his bank balance and that the other makes the land his bank.

The Curious Traveller (1950:29-38)

Stubble Burning

THERE were three infallible means of identifying at sight the brand of the combine on the September fields of Wiltshire, which looked to me like a vast empty camp of stockless mechanised farmers. One was by the policy of scorched earth, namely by burning the straw after it. The second was by the broad striping of green lanes along the entire length or breadth of a stubble-field; these were where the grain dropped by the machine had sprouted. I learned from Hosking after I had returned home that he had observed the same phenomenon in Oxon, Bucks, Suffolk and Cambridgeshire. The third was by the *farouche* appearance of the field itself.

To take the last first, I have never in all my wanderings for thirty years over the face of England seen cornfields like those of 1944 in Wiltshire, one of the most progressive and highly mechanised agricultural counties in England. Here is an example, one of scores. The field was a large hedgeless one and looked as though the troop of Bacchantes in Keats's poem had charged madly into it, trampling down the straw in the ecstasy of the vine and biting off the ears. It is said that one advantage of a combined field is that it makes good cover for partridges; here, as one member of our party remarked, was good cover for a rhinoceros.

Parting with some effort the tangled clumps of straw, we could see the hard ground strewn with countless grain and weed-seeds. On the edge of the field stood a caterpillar tractor ticking over and attached to a five-furrow plough. So the ploughman told us (and he had had 18 years' experience of ploughs). We had to take him at his word because most of the plough and all of the shares were invisible. They were wrapped up in straw as though the plough had come by post in a straw packing, and the ploughman was busy with a pitchfork unpacking it. It had been impossible to burn the straw because there was an ammunition dump close by.

Elsewhere, there were no such obstacles. One of the purest landscapes and richest greensand soils in England is the edge of the Vale of Pewsey between Alton Priors and Allington, backed by the long line and masterly modelling in col, headland and saddleback of the Marlborough Downs. The very cottages of timber, brick and stone respond to the simple and economised grandeur of the natural composition by touches of special elegance – curved struts, rhythmical sweep of thatch and tile-headed oriel windows. The distant spire of Bishops Cannings Church serenely clears its ramp of trees. Another religion as organically impresses itself upon the landscape. The long barrow of Adam's Grave forms one slope of a green conical mound and below it the land falls away to a wide terrace supported by flying buttresses on two sides. The natural takes the form of man-made architecture, the man-made had conformed to but enhanced the natural design. Lofty Tan Hill directly opposite held memories of the great sheep-fairs thereon, stressing the particularity of this place as great in husbandry as in beauty.

But now the land was sheepless and stockless, while the green symmetry of the flowing Downs confronted a black smudge, an area of burnt-out desolation. For to burn the straw gives a saving of 3s. 6d. per acre of cost. A short time before, it had been a barley field, as we could tell by the strips of green made by the self-sown grain and the myriads of kernels lying on the ground, each one with a black spot on it from the burning of the straw. The disgrace of this field was dramatised by the grace of its setting.

It is said of those who advocate the cultivation of the earth by good husbandry that they are "warped by nostalgia" cramped by traditionalism," that in their desire to "keep farming free of all taint of commercialism and industrialism," they "grope about in a sort of mediaeval twilight towards a mystical state of which the symbols are the maypole and the night-soil cart." The remedy, it is argued, for this mystical mugwumpery is the "rationalisation" and "standardisation" of "the agricultural industry" in the interests (viz.: the vested interests) of international trade. Food is not grown to satisfy people's needs but mass-produced as a factor in the exchange-value of competing for a favourable trade balance. Here was this "rationalised" (viz.: labourless) farming in visible action or what has been called "combine and matches" farming. On the Downs above us, the Bronze Age colonists of Avebury had cultivated barley with the digging stick. And this primitive scratching was superior farming practice to what I saw below the Downs 2000 years later.

The truth is, as Hosking put it as an appendage to his comments on the tractor, that the combine cannot take crops too good for it. It is simply a method by which the inferior farmer with thin crops can beat the good one with heavy ones by cheapening the process. He compared the combine farmer with a roundsman who buys a faster car to do a quicker round and so sell more goods. It is a process impossible with horse-farming or spade-cultivation because the only difference there is in the quality of the work. If the combine were improved in an attempt to stem the waste, it would become even more cumbersome and unmanageable than it is already. Its only advantage in that it allows the grain to ripen in the standing ear; its only excusable use is for short-stemmed barley and we had seen what that was like under Tan Hill. The potash from the ashes is the merest short-term stop-gap against the drain of fertility.

Mechanisation of the land has, in fact been adopted to fit into a modern economics based on a purely quantitative production. Yet, as we had seen it, it had defeated its own short-sighted ends. "The folds shall be full of sheep; the valleys also shall stand so thick with corn that they shall laugh and sing." But not by power-farming and "efficiency." If the modern combine-farmer answered his critics in some such terms as these: "combine farming is unquestionably wasteful both in grain and in straw; it is using up fertility because the chemical bag is the only fertilizer to replace the drain on organic plant-foods; it is expediency farming. But we cannot help it. Nearly all our labour has been taken from us; our livestock has been depleted by a bad policy, and good husbandry, through no fault of our own, has been at a discount" – if such were his plea, the edge of criticism would be turned. But what he actually does say is that he must "move with the times," as though the waste and expedience were not only inevitable now but part of a future "prosperity". Thus he allies himself to a mentality which thinks of human beings as "mobile males," "immobile females," "dilutees" or just "bodies." He becomes the victim of a habit of mind caricatured in the following, taken from the pages of a farming journal:

> "Mechanisation of farming has, to a large extent, stopped short at replacing the horse. There are still more farm employees working with their arms and legs than there are on tractors, and this is likely to be uneconomic in future."

Unfortunately, the natives of this armless and legless Utopia will, by such farming methods, have nothing to eat.

The Wisdom of the Fields (1945:210-13)

Friend Sykes

SOME years ago I lunched at the Farmers' Club with Sir Albert Howard and a farmer convert of his, Mr Friend Sykes. He farms Chantry, high up on that wonderful piece of chalk country where the land falls steeply away from Chute Causeway into a huge green cauldron whose downward-rushing sides are clothed in summer with ragwort, cloth-of-gold to the sightseer but to the cultivater the badge of a beggared land. This is Hippiscombe and opposite rises Haydown Hill, chequered with the low banks of the Celtic tilth, and, with the Celtic citadel of Fosbury, connected by a track with Tidcombe Long Barrow, thrusting out its revetment from the woods behind. Eagle's country, this noble preface to the watershed of Avon, Test and Itchen, but the ragwort is indicative of something else than visual glory. There is hardly poorer farmland anywhere else on downland, and when Mr Sykes first took over Chantry the 10,000 rabbits killed were such starvelings that they could be neither marketed nor locally consumed. Moreover, the farm was riddled with disease, and the crops were barely worth harvesting.

Look upon this picture and on this! I have beside me a record, written by Mr Sykes, of the condition of his farm of 750 acres to-day, when for thirteen years he has been consistently applying the Howard doctrine of humus and nothing but humus. Tuberculosis, mastitis, contagious abortion, Johnes' disease and the others have all vanished; the 2,000 yearly visitors to Chantry remark upon the bloom of health on the stock; the cows keep up high yields on grass alone, and the grass, cereal and root crops continue year by year to increase their output per acre. For instance, the milk-yield was 8,000 gallons more in 1948 than in 1947 without an increase of the number of cows in milk, while the butter-fat and solids increment was 5 per cent. In spite of a drought even more tenacious than that of 1921, the 1949 yields show a yet further increase in quality no less than in quantity.

It is, therefore, of cardinal interest to learn by what methods Mr Sykes has achieved such spectacular results, and without recourse to any of the fashionable stimulants so deceptively called "fertilizers." First and foremost, by mixed farming, one of the great heresies of industrial agriculture. The farm carries an attested Guernsey herd, one of Galloway beef-cattle, a flock of half-bred ewes crossed by a Suffolk ram which averages two lambs to a ewe for the whole flock, poultry and eighteen thoroughbred horses, with six mares all in foal this year. There is no sterility, the bane of modern farming, in any of the stock. The bullocks are yarded and convert straw into muck in the old-fashioned way. But all the waste vegetation of the farm – hedge trimmings, roadside cuttings, ditch cleanings, bracken and leaf-fall – is also collected and mixed with the muck to form compost, and of this 1,000 tons are made yearly. Not nearly enough on a farm of 750 acres, and so Mr Sykes makes the backbone of his farming the deep-rooting four-year ley, ploughed up at the end of its term, so that the 25 tons of dung and urine deposited by the mixed grazing may perform the office of activator for turning the sod into humus. This, of course, is sheet-composting, and it is important to note that the ley-mixture sown at Chantry is composed of no fewer than fifteen different grasses, clovers, green crops like lucerne and sainfoin, and deep-rooting herbs like burnet, chicory and yarrow that act as subsoilers for bringing up valuable plant foods into the topsoil.

The modern ley system is a failure for the reasons that rarely are the mixtures more complex than clovers and ryegrass, make little use of deep-rooting plants and are seldom long in term. In other words, they pay scanty attention to the very mixed ingredients of natural pastures, which Elliot so strongly advocated in his Clifton Park system. Here, then, is an example of a farm which makes handsome profits, produces heavy yields, supports abundant mixed livestock, has been vitalized from extremely poor land and has enormously increased its fertility by the intelligent and exclusive use of humus. Humus has been the economy of the farm, humus and nothing else, for thirteen years. The farmers say they cannot afford to do likewise; the State subsidizes an entirely different concept of agriculture; the scientists bring every kind of pressure to bear upon the adoption of their laboratory farming, and humus-farming is cried down as "primitive" and outmoded. Only a tiny minority has the crushing answer – it delivers the goods.

The Faith of a Fieldsman (1951:119-21)

Compost

MODERN civilization with all the world as its handmaid is as barbarous in the food it eats as it is barbarous in its treatment of the earth that grows it. A civilization devoid of the sense of wholeness of living, such as the smallest garden properly cultivated displays, is beneath all its chromium plating uncivilized.

Of that wholeness the compost heap is the linchpin. Without this generating station of humus, the whole edifice of self-sufficiency collapses, and with its collapse go all the food-realities. Fetch your food from the garden as fresh as you please; if it is not grown from humus, that is to say, out of the flesh and blood of the garden itself (with a bit of aid from over the hedge), it might almost as well have come out of Covent Garden. The vital link is missing. I once heard a B.B.C. talk by one of the Rothamstedians to children. It advocated certain kinds of fertilizers they should use in their gardens for different types of plant. There may or may not be something to be said for the use of artificials on the large-scale mechanised farms (our modern *latifundia*) – and from the point of view of organic agriculture nothing at all can be said for it. But there is not the faintest shadow of an excuse for them in any self-respecting garden or small holding over and beyond that of swelling the profits of the enormous interests whose business it is to see that organic husbandry shall become obsolete. It is not only completely unnecessary for children to buy artificials for their gardens but positively corrupting to their general mentality to be instructed in their use. Why? Because it gives them a false conception of the kinds of laws, intricate and exquisitely disciplined, which nature employs in maintaining and repairing the architecture of plant-life. The inorganic plays the lowest part of all in this great drama of interdependent and interacting life-in-death and death-in-life. But the children would go away from that lecture with the idea in their budding minds that chemistry – a clever trickster – was the real hero of the play.

I do not say that the man giving the talk intended to convey that impression, but that must have been the effect produced. It could not fail to distort and falsify the child-sense of the wonder and mystery of organic life. Every child should have a religious view of life, and this is a religious view. It is an abomination that a child should believe chemistry to be the dominant power in life. Leave it to the adults to assume that nature is an atomic force that sets us "problems" in the manipulation of matter and energy. The only possible inference to be drawn is that the actual life of nature, if it can be said to exist, is an irrelevant epiphenomenon, a kind of froth exuded out of the complicated energies of heat, motion, light, radiation, electrons. It is singular indeed that, while materialism as a philosophy of human life is now utterly discredited, it should be assumed to be valid for plant-life. If human beings cannot be kept alive on chemicals, how should plants? There can be no real doubt that plants can suffer from deficiency diseases, the effects of malnutrition like our own. Observers *in the field* have noted that insects, virus and other infections rarely attack other than weak and leave virile ones alone. They are in fact nature's sharp warning that her plant is being improperly fed. Not only does this vicious doctrine teach the children to see nature "as in a glass darkly," but to be lazy and parasitic. They learn how to take the short cut. Chemical pellets save them from observing the dietetic preferences of animals, how, as experimenters have now proved, mice, rats, rabbits, poultry, sheep and cattle will always choose foods organically to inorganically manured. They grow up in the magical superstition (like the sacred congealed blood of St. Januarius) that what makes the garden grow is something out of a bottle, a sack, or a tin. Science originally made war on superstition and has now enthroned its own. No wonder that the majority of the nation's children believes that milk comes out of a pasteurised milk-bottle. Educate them from magic to truth. It comes out of a cow . . .

The principle of the compost heap is both the principle of nature and the principle of wholeness. Thus, orthodox science, in opposing it as it has repeatedly done, has progressed from the interpretation of nature to setting up systems as hostile to nature as are its specialisations to wholenesss of living and thinking. The true but heretical science of earth-culture is that of Sir Albert Howard's now celebrated Indore Process. This is simply an analytical examination of nature's own methods of regenerating her forests by means of the mixed humus of the woodland floor, and the constructive adaptation of natural humus manufacture through bacterial energy to man's own need for regenerating the soils which he plants and cultivates for his own purposes.

Like most of the more genuine discoveries of modern days, its roots are in the past, as far back indeed as Tusser's *Five Hundred*

Points of Good Husbandry. An Arabian wrote a treatise on the properties of compost in the 10th century. According to contemporary testimony, the Islanders of Barra made compost in 1794, and the Aran Islanders made their barren rock fruitful by the use of it. John Evelyn made a prolonged study of it in *Terra,* and Speed in *Adam out of Eden*. . . . Cobbett recommended compost with his usual lucidity of detail ("A great deal more is done by the fermentation of manures than people generally imagine"), and Dr. G. V. Poore, a consulting physician to many London hospitals, born a century ago, advocated the saving of city wastes and the use of night soil for conversion into humus in his *Essays on Rural Hygiene* and other works. The traditional practice of the French is to make "terreau," very well rotted dung mixed with soil and frequently turned. This produces 4 market garden crops between January and July on the same soil every year. The Chinese have, of course, been compost-makers for centuries and for 4000 years have held the belief that every person with access to land can support himself from *his own* waste products. This is one good reason why they have preserved their civilized structure from the collapse that has befallen other cultures less tough than theirs. These systems knew from traditional practice the processes and results of composting. But, if Dr. Poore's pioneer work be excepted, they were unaware of what actually took place in the translation of organic and vegetable residues into plant-food. To establish this on a scientific basis has been the work of Sir Albert Howard in his *Agricultural Testament,* and, to a lesser extent and with some occult suggestion, of Rudolf Steiner and Dr. Pfeiffer. To these honourable names should be added those of the botanist, Prof. Bottomley, who invented the formidable titles of "auxinomes" for vitamins of plant life which are found only in organic substances, and of Prof. Gilbert Fowler, an erudite and zealous exponent of the "rule of return."

Hamlet remarked to his mother: – "Spread the compost on the weeds to make them ranker." But what makes plant-growth rank is not compost but the use of chemicals as a plant-food rather than a plant-stimulant, just as the face of a man who drinks too much whisky becomes puffed and mottled. A closer analogy is perhaps the use of drugs for insomnia. In a neuropathic condition, a sleeping draught may restore the normal rhythms of the patient's metabolism; if persisted in, it destroys the very sleep it was its purpose to induce. Humus is the daily bread of the living soil. If for one reason or another its condition is impaired, a phosphatic or nitrogenous fillip may well repair it; to continue such treatment is, when it asks for bread, to give it a stone. In a garden where I was staying, the cabbages, artificially manured, were twice the size of my own. But they were all without hearts, whereas mine, compost-fed, had all dense and tight hearts . . .

It is difficult to spare quantities of soil out of a garden whose every square foot is in use for one purpose or another. Fortunately, I have a good dyke outside my eastern hedgerow whose oozy bottom makes one pinguid layer and nettles another. I cannot praise the nettle too highly. There is a yeoman I know whose one reason for buying his farm was the nettles on it; if nettles grow on it, he said, anything will. Good for making paper and linen, good for brewing nettle-beer, good, as Shakespeare knew, for growing strawberries under them, good too for guarding plums from wasps and good again for ripening the plums on beds of them, good for raising the temperature of the stack, good for salads when young, good for the whitethroat or nettle-creeper, a fort of a million spears for her fragile cradle and very good for the compost stack from their potential wealth of calcium and potash.

All the household wastes I also use with the exception of those that go to the livestock or have printer's ink upon them. And nothing that grows in the garden but finds here its last home and bed, to feed the living by the due end and course of nature and by the mysterious cycle of life, decay and renewal. Even the prunings and hedge-clippings I strip of their leaves in death's service for life. Stout woody stems like those of Brussels sprouts and other Brassicas I either crush with a maul or use for foundation or aeration. Urine swells the offerings, though I have not yet had the courage or the means to experiment with night-soil, which Cobbett advocated for its uses in fermentation. Luckily, I am surrounded by small farmers and so can make use of horse and cow manure, apart from the ample contributions of my own livestock. For mixed organic manures are the best.

Four stakes are thrust into the heart of the fermenting and decomposing mass for the air to sweeten and aerate it. I turn and rebuild it, keeping the green stuff on the inside, when time permits. When the complex processes of metabolism are completed, the result is a rich currant cake, highly compressed but breaking down into a friable mould that will go through a sieve, odourless and of the consistency of potting soil. The only "chemicals" I use are lime and potash from the ashes of my peat and wood fire or from burnings of hedge-cuttings and prunings. But it behoves the composter to be careful with lime. It tends to dry the stack.

Many gardeners who adhere to the principle of the compost heap use sulphate of ammonia or calcium cyanomide for hastening the processes of fermentation and decomposition. Even the excellent No. 7 Ministry of Agriculture Dig for Victory leaflet approves of the use of chemical fertilizer on the compost midden. This is characteristic of the modern superstitious magic of the Djinn in the bottle called by some heavy-armoured name in bastard Greco-Latin. But all it really means is the modern passion for short cuts and dodging nature. These chemicals certainly dislocate the balance of the stack, as does excessive dung, by encouraging the anaerobic at the expense of the humus-forming bacteria and fungi and so releasing the nitrogen into the air as ammonia. For, though humus contains all the chemical ingredients in proper balance that are now applied piecemeal to the soil from factory by-products, its organic action is not chemical at all. The real function of humus is "to stimulate the activity of soil fungi" on which the plant-rootlets depend for their vitality and immunity to disease. These fungi "feed on the humus in the soil and it is the product of their metabolism which is of such vital importance to the complete and balanced nourishment of the plant." Humus is a substance that comes from the once living.

This discovery goes to the heart of the illusion that chemicals feed plants and explains why disease always accompanies the use of chemicals without humus. The activities of these soil fungi *on which all life ultimately depends* are inhibited. Chemical fertilizers also affect the closeness of texture in the ultimate mould and so its moisture-retaining qualities. Their brusque interference with the natural workshop (like an inspector bursting in on a factory and hustling the work-people) hardly ever accomplishes its purpose. Fortnightly applications of lawn-mowings most effectively expedite the formation of humus. The chemical changes are brought into action by intensification of heat and it is impossible to put one's hand into the stack after the lawn-mownings are on. Come a shower and it smokes and will sink two or three feet in as many days. When a stack is thus treated according to the prescription of nature, it is quite free of flies and odour and is ready for use in three or four months, rather longer in winter and if the weather is very wet or dry. If it is made in a workmanlike fashion, all weed-seeds are killed, all spores of virus and all "pestilence-stricken" leaves. When weed-seeds germinate (as they sometimes do), the fault is in the composter, not the compost.

I have already described the almost miraculous effect of a well-made compost upon the fertility of the soil and the vigour of plant-life. These in their turn ensure freedom from disease and the good health of men and animals that feed on it, while the vitality of the plant is expressed in the palatability of the meals made from it.

What could be duller than minding a machine in automatic repetitions which call for practice rather than skill? In their products the worker has neither share nor interest and over their processes no control. But there is necessary drudgery even in the craftsmanship of gardening. This is lightened by composting in three ways. First, by the extraordinary interest of directing and manipulating a garden to support itself. Secondly, by the elimination of all the niggling, highly specialised and fragmented labour of biochemical analysis in studying the inorganic requirements of your plants. For the enormous benefit of humus manufactured by composting is that it is its own laboratory. It supplies all the oxidation, all the chemical salts necessary, the nitrates, the phosphates and potash compounds, which are absorbed by the roots and carried up in the sap to the green leaves. Thirdly, and this follows naturally upon No. 2, the toils of weeding are carried upon a new plane.

Weeding considered *per se* as a pursuit has little to recommend it. Yet weed you must. Charlock is the host of the flea-beetle and cruciferous weeds are of the slime fungus that brings club-root to cabbages. Weeds compete with food-plants and flowers for the supply of nutrients, especially nitrogen, and no gardener worthy the name can bear to look upon

> "an unweeded garden
> That grows to seed; things rank and gross in nature
> Possess it merely."

To an authentic owner nothing is more unpleasing than a dirty bed. But if the weeds go to the compost heap, the process of getting them there acquires a new significance, both practical and psychological. The weeder is acting one of the rhythms of nature but at the same time controlling it. And as he weeds he educates himself. For weeds are indications of soil deficiencies, and by a curious paradox – and life is a perpetual paradox – weeds in a garden are rich in the particular minerals that soil lacks. The more vigorous certain types grow in a particular area of soil, the more lacking that soil is in those needs. Bad drainage, for instance, is betrayed by marestail, ranunculus, mosses, meadow sweet and other plants. Acidity is advertised by dandelion, plantain, daisy, dock, self-heal, bents, sorrel and others. Nitrogen deficiency is registered by the presence of the nitrogen-fixers – clovers and vetches. It is an interesting fact that nitrogen-fixing legumes like clover, if artificially overfed with nitrogen fertilisers, become consumers instead of producers of nitrogen. If the weeder glances at the nodules on their root-hairs, he can tell by their presence or absence whether his soil is rich or poor in nitrogen.

Thus, the weed extracts from the soil exactly what the soil needs, and the prevalence of one type of weed over another is a pointer to reading the soil like a manuscript. The absence of boron, cobalt, manganese, potassium, sulphur, phosphoric acid, sodium, zinc, iodine, silicon, copper, iron, or some other essential or trace-element in minute quantities becomes a language to be deciphered. The profound error of modern science has been in separating minerals and vitamins *from the plant,* whether in health as food for beasts and men or in decay as food for the soil. This is the root of the matter.

But let not the student be dismayed. The compost heap shoulders all the burden and he can lean back on nature's erudition. She is the librarian. Trundling his barrow filled with weeds, the weeder flings the whole bundle on to the stack. In time the fungi and bacteria, on whose activities the whole sensible world depends, reduce them to mould which in the spring or autumn is forked into the soil or laid on as a top dressing. Thus are restored those very elements whose shortage of supply was announced by the living weeds. The compost heap is the most precise of chemists and its prescriptions are impeccable. How then can the weeder regard his task any more with distaste and weariness? Bent to his toil, he is also the student bent to his book. He bows his head to the wisdom and economy of nature, and, acquiring this knowledge, perhaps offers a silent prayer to the Creator both of nature and himself. For the compost heap corrects the fallibility of his judgment and redeems a prosaic task of its pedantry and tedium. It gives him learning without tears.

There is yet another thing he learns – and that is the real obstacle to making this garden Horn Book a lesson of universal application. For, quite apart from the mutually contributory (scientific jargon calls it "symbiotic!") relation between plant and fungi of humus which penetrates the root-cells of the plants by threads rich in protein ("the mycorrhizal association"), the humus is itself mineralised by the soil bacteria. Thus NPK *and* protein are prepared by nature, the minerals seeing to the size and growing power of the crops (the yield) and the protein the quality. If, that is to say, the preparation of humus was extended from gardens to the whole earth, the vested interests in artificials, poisons and soil-dopes of all kinds would cease to be . . .

Another benefaction of compost is its conservation of moisture in the soil, and what this has meant during the present cycle of drought-years cannot be conveyed on paper. When a period of drought has set in, a depression means nothing, the appearance of the sky means nothing, the direction of the wind means nothing. Rain-clouds look as promising as a telegram of confidence, sombre, dragonish and big-bellied. But they are no more than fancy-dress clouds, impotent and inglorious with no fatness in them to drop, no balm to distil upon the thirsting fields. How often in my region have I not seen the roots of the peas turn white, the potatoes flop, the turnip-leaves go bleached and yellow, the hard fruit lose its hold for lack of sap, the soft fruits look as woody as they prove to the taste, lawns turn sere and the soil yawns into cracks that will take your foot! The drought-winds of 1943 shrivelled the very leaves off the plants three months before the fall and shrank the kernels of the dry-loving wheat to half their normal size. Often I have had to rely upon mist and dew for rain in essential planting out . . . Yet since I methodized my composting, my actual losses from drought have been infinitesimal. The flowers act as though in a film but the produce has never sensibly lessened. In 1943, for instance, the spring drought lasted for three months and the summer one for another three with intervals of only tea-cup rains. But, except for surface-rooters, I never, even in this heart-breaking year for the gardener, used the watering pot. My soil was even fit for light planting at the end of July, 1943, after a drought, broken only by showers, which lasted from mid-February to mid-October. The morning dews helped but the real stay of plant-life was compost. The main damage was caused by birds, who, desperate not so much from thirst as hunger owing to a soil like a pavement, voraciously attacked the hard fruit and many crops.

Contrast this successful resistance to drought with the conditions in the 1943 East Anglian drought, as severe as in my neighbourhood. Complaint was loud among the farmers and the crops suffered acutely from the lack of rain. Why? For the obvious reason that humus was lacking. The high traditions of East Anglian farming are giving way to the chemical complex. The consequence is that soil can no longer resist drought as it used to when those traditions were much more active. What the farmers were complaining of was not drought but soil-exhaustion.

I also have shrewd notion that composted humus regulates the temperature of the soil, making it cooler in summer and warmer in winter, like a roof of thatch on a house. Humus, too, has important properties of binding the soil-particles and preventing them from being blown away as in the dust-bowl areas.

Compost exercises, too, a transforming effect upon the texture of the soil. Even the colour of my own, naturally reddish, has in course of time turned almost black. No matter how beaten down by heavy autumnal rains, it yields at once to the fork as a flakey crumbled tilth that would warm any gardener's heart and give him as much to look at as a bank of flowers. The compressed clayey patches are loosened and the more sandy areas tightened up. The effect of compost is also to enrich the colouring and add lustre to the flower-heads of plants. Again, it ensures germination. This is particularly noticeable with onions and shallots. In one year and sowing the same amount of seed, I increased the onion crop by a bushel and a half, the shallot by a peck. Lastly, aeration to prevent capillary action in a dry spell can be left for much longer periods than if the soil had been fed with farmyard manure alone.

This Plot of Earth (1944:99-108)

Earthworm Gardening

IT chanced that I was chairman to a meeting organized by the local education authorities, and a long address was given by a local gardener upon his particular methods of growing flowers, fruit and vegetables. The lecturer was unknown to me and I had no idea of what he was going to say. But since the word "modern" appeared on the title, I expected to hear all about the latest thing in insecticides, spraying machines, sterilized soil, growing plants in water and chemical solutions (hydroponics is the technical term, is it not?) and what not. In fact, I wondered what on earth I was going to say in introducing the speaker, since my faith in the laboratory gardener is hardly a fervid one. However, the word "modern" is beginning to mean the reverse of what it meant a few years ago, which was simply industrial. Industrial methods, whether in garden, farm, forest or any type of land, have proved so catastrophic that, by an ironical paradox, the word "modern" is being interpreted nowadays as the scientific study of traditional cultivation and the discovery of new processes in harmony with it. So it proved in this instance, for the speaker told his audience that he did not cultivate his garden at all, but bred earthworms to do his cultivations for him.

What he does is to collect some 500 earthworms into a wooden box about the size of a sugar crate and filled with rich, well-moistened, composted soil. In this the worms (not the big migrant lob worms but the smaller and stay-at-home type) lay countless eggs, and in three weeks' time the soil is emptied out of the boxes and mounded up into a cone, so that the worms make their way down to the base. The eggs are then easily collected and deposited into a compost heap made up not of lush spring waste, which would burn them from the heat it generates, but a cooler one made up of the new soil taken from it. When this second heap is ready for use, it is spread over the garden, worms and all, and these proceed not only to make as much topsoil in a year as undisturbed nature makes in about 1,000 years, but to aerate it, to give every facility to aerobic bacteria and to fetch up the valuable minerals from the subsoil. In other words, this "modern" gardener does no digging at all; his spade is the mysterious lowly, secretive earthworm about which poets loved to philosophize as the final disposer of our mortal life, but to which Darwin thought it worth while to devote a treatise.

This was all highly original and stimulating, not to say provocative.

But the proof of the pudding is in the eating, and I conceived a great curiosity to go and see the garden to which these novel and heretical methods had been applied. And go I did. It lies, four acres of it, 600 feet up, close to the Oxon-Bucks border where the Chilterns display in miniature the characteristics of a well-wooded intricate downland seamed with winding little dry combes that give it an aspect of great charm and diversity. And I am bound to say that the garden lived up to its delightful environment. It was not long after the terrifying tornado which practically destroyed the whole of the most promising apple crop I have ever had. This garden showed no sign of the calamity; on the contrary, it looked as though invisible walls had shut out all the storms and stresses, the vexations and frustrations, the failures and anxieties which are the common experience of the gardener. The lustiness and exuberance with which the flowers and fruits of this garden were growing were something not to be forgotten. I have always rather prided myself on my strawberries, but those in this garden were the finest plants I had ever seen. Weeds and slugs were defeated by a liberal use of sawdust, and of all the Egyptian plagues that assail the gardener and farmer I saw no trace. Wordsworth has often been ridiculed for declaring that every flower "enjoys the air it breathes"; the scoffer here might well have been mute.

I have no wish to dogmatize on what I saw, but it was obvious, at any rate, that this gardener had not been talking in the air. So far as his own garden was concerned, the facts gave his thesis overwhelming support. That does not necessarily mean that earthworm-culture and "no digging" are the only key to the maximum of fruitfulness. But it surely does mean that we have much more to unlearn than learn in our understanding of nature, and that the organic way, however it be interpreted, is, in the long run, the only and the best way.

The Faith of a Fieldsman (1951:75-7)

D.N.O.C.

ON a plot of land adjacent to my orchard and garden and rented from a farmer, I have, for fifteen years in succession, grown a crop of potatoes. This plot is not more than about 100 yds. long and 20 broad. It also grows a large variety of other vegetables, is a nursery for flowers, and carries soft-fruit bushes and a few trees that are an overplus from the orchard. Since these potatoes take up nearly half the ground and the bushes and trees are more or less in permanent quarters, it is clear that, even if I wanted to, I have little scope to shift these potatoes from one area of the plot to another. They are in part, that is to say, grown on the same ground year after year. The only trouble I have ever had from these potatoes is that in wet summers I lose a very small percentage from slugs. They continue to be extremely prolific; have never had the smallest trace of any disease in very wet seasons except an occasional touch of blight which does no harm to the crop; are particularly robust this year – their sixteenth; are incomparably better flavoured than the potatoes one eats in restaurants; and have never been fed with anything but compost which consists not a little in their own decomposed haulms. The same is true of the rest of the produce on this plot. From the vast cornucopia of pests and diseases which afflict farm-crops I have been immune except for occasional attacks of flea beetle, which I counter with Derris powder. The enormous ramifying vested interests which make their profits out of sprays, fertilizers, insecticides and chemical nostra of one kind and another have never received a penny from me, while, at the same time, the crops I grow are superior in every respect – output, health, flavour, quality *and* quantity – to those I see so often on farm-lands.

Now I do not write this as a cock crowing on his own dunghill. There are plenty of other gardens that pursue the same methods as I do, and with the same results, and my point is that such gardens (the majority, I should say, superior to mine because the time I have for gardening is very limited) are so far ahead of the average farm in productiveness, quality and intensive husbandry that there is really no comparison between them. The plants in such gardens give of their best because they receive what nature demands for the vitality and fruitfulness of her creatures; the plants in such farms are permanent patients in a hospital, kept in a sickly apology for life by steadily increasing ministrations of drugs, disinfectants and pick-meups. They are invalids from birth (emerging from their mercury coating) and pass on their depression to their human consumers, since man is to a large extent what he eats.

I am led to this very obvious contrast between the healthy garden and the disease-ridden farmlands by a long letter I have received from a farmer calling my attention to the lethal possibilities of the patent new man-killing poison which has recently made news – D.N.O.C. "It is not for me", he writes, "to enlarge upon the damage to bees and birds, the loss of pigs littered down with straw from sprayed crops, the destruction of soil organisms, and the unknown changes in food composition which may occur as a result of this supposedly cheap method of eradicating weeds." Indeed, it is hardly necessary to stress the repercussions of this horrible venom upon farms, when it can kill a man and turn him yellow in the act of spraying without the armour of a rubber suit. The use of this pernicious poison upon weeds in place of the honest hoe is a perfect example of the utter bankruptcy of good husbandry, and its displacement by the equally deadly one-track-mindedness of modern agricultural science, in complete oblivion of the fact that pests and diseases continue to multiply in spite of it. I remember the time when moneyed and official pressure regarded one spraying per year as adequate. Now it is at least seven times. What a confession of failure!

For this science to confess itself wrong, as common sense cannot do other than know it to be, would be a supreme miracle. Yet this evil fatality of poisoning the fields with ever more potent distillations from the laboratory as short cuts and financial expedients to avoid man-power and proper cultivations, is already having such vicious results that reason insists upon breaking loose it. Is there nothing that will teach this fraudulent type of science to examine its own pretensions? I do not know, but should there be any future prospect of our recovering our senses, there is always the garden to point out the true path.

The Faith of a Fieldsman (1951:132-4)

Organic Pest Control

WHAT remains of one of my legs having made an effective protest against the great heat, there was nothing I could do but, as in Ralph Hodgson's *Song of Honour,* "stare into the sky." A brisk note sounded and a blue-tit alighted on the telegraph wire before disappearing into the nest within my guttering. I am pretty sure that this bird and its mate were the same pair that stripped the fruit-buds from two of my cordon pear-trees in April and occasioned me hours of labour in black-cottoning the rest. In my compulsory inactivity, I watched the pair coming and going for some hours. Each bird was feeding the young roughly once every minute and a half or forty times an hour. Thus there were eighty visits to the nest per hour. But natural life escapes the yard-stick, and occasionally one bird would be away for a quarter or even half a minute longer than usual, while both of them used to pause for a few seconds on the wire in order to pant. I will therefore, allow a quarter of an hour to each bird to account for these irregularities in the routine. Thus it is a conservative estimate to reduce these feedings from eighty per hour for the pair to forty. The birds continued to feed their young at this for the whole of the time I was watching them, about three hours. Counting the hours of daylight as eighteen at this time of the year, the paper calculation would be over two thousand feedings per day. Deducting a quarter of these for rests, distractions, personal meals, slowings up for weariness and occasionally longer excursions for prey, and I arrive at an all-round estimate of fifteen hundred meals for the young every day, or ten thousand or so a week.

While I was watching them, they certainly never crossed the boundaries of my land, and each bird pursued exactly the same course on each journey to and fro. One went straight down south to the orchard, the other east by south to the cordon and espalier trees that run down towards the orchard. It is fairly safe to assume that the birds did not during the day swing their helms to any great degree port or starboard of these fixed points.

That brings me to the nature of the cargo. Since the birds were some twelve feet directly above me and always alighted at the same place with their beaks facing me and the prey hanging from them against the blue sky, I could comfortably ascertain what it was on each visit to the nest. It consisted entirely and exclusively of one grub of the Apple Blossom Weevil and one maggot of the Apple Sawfly, never more than one, one hundred and twenty of them as I watched. The weevil, when immature, attacks apple buds at their pink stage and makes a brown casing over the corolla. Having destroyed the potential apple, it burrows, when mature, through a lateral hole in the bud and makes skeletons of the young leaves. In a prolific year it may do useful service in thinning out; in a year like the present when the drop is heavy from lack of bees, it exacts a taxation of the crop to be almost as severe as the regulated prices of Whitehall to follow. The sawfly lays its eggs in the calyx of the bud and the hatched maggot bores into the centre of the fruitlet, moving from apple to apple, and, it has been estimated, destroying 40 per cent. of the crop before, satiated, it goes to ground to emerge as the fly the following April. Since the pair of blue-tits was a cost-free labour force, ridding my trees of such pests at the rate of forty or fifty an hour, I could more than afford them their wages in advance when they stripped my two pear-trees in March. Had I sprayed the trees with D.D.T. for the weevils and arsenate of lead for the sawflies, I should have destroyed the insect predators as well as the grubs, poisoned the soil, killed many earthworms, spent a good deal of money and labour and presumably lost the services of the blue-tits.

An Englishman's Year (1948:182-3)

Jackman's Nurseries

I was standing in the churchyard of Woking with one of those very rare birds, a native of Surrey. He pointed out to me a patinated gravestone which stood outside the Victorian period and belonged to one where shape and proportion, good lettering and a natural sense of beauty had not yet been sacrificed to Cobdenism and Benthamism. That, he said, is my great-grandmother's grave. We walked over to the church porch. It has the original Norman west doorway built about 1080, iron-braced and strapped and hung with finely wrought iron hinges. These are still in place, though the door is hung to-day with modern screwed-in hinges. It is profusely decorated with iron ornaments which at the same time serve the structural purpose of holding the half-inch planks in place – a cross, a saltire, an eight-legged spider and studdings of horseshoe nails doubtless smelted in charcoal from the Andredsweald oaks. The tower itself is, so to speak, in two halves, the lower one being Norman and built of stone, flint and thin wafers of brick. The brick must have been, of course, much later than the flint and stone work, but was recognisable as characteristic of the brick work in the old cottages of Surrey. In the churchyard were graves of brick, semi-cylindrical and of the same shape although smaller than the curved sarcophagi of many Cotswold tombs. It seemed incredible that I was seeing these things in the very heart of suburban Surrey, if Suburbia can be said to have a core or hub.

But most remarkable of all was that the man who belonged by family and original inheritance to this ancient Surrey with its continuity stretching back to Saxon times and indeed much earlier, was the managing director of one of the most commercially prosperous and celebrated Nurseries in Britain. He is Mr. Rowland Jackman, and I was spending the day with him in order to see under his guidance Nurseries whose plants have become a household word everywhere.

There are fifty acres of them and another hundred of plantations, seventy of them let off to a farmer, the whole area employing no fewer than ninety workers. Mr. Jackman told me that among them are factory workers, men and women, who could not endure the dullness, repetitions and soul-degrading uniformity of factory life, and they of course would be the best types. The nurserymen, on the contrary, constantly change over from one occupation to another, potting up plants, pricking out seedlings, working in the glass houses, budding roses, grafting, cultivating, hoeing, making new concrete frames, forking them over with wire rakes made on the estate, together with a multiplicity of other jobs. For the whole principle of these Nurseries is to make them as self-supporting as possible.

A foreman, for instance, whom I saw constructing a frame, is also a heather-thatcher, and one of the barns is so thatched, the material coming from the neighbouring commons and the technique being a surviving example of an almost defunct Surrey craft. The tied heather was held in place by sweet chestnut laths, while the external surface was pegged down by three parallel rows of crossed hazel rods. The admirable timbering of this barn was the work of a young

Quaker who does the repair work on the small machines used in the gardens, the rotary hoe, for instance, besides carpentering, smithying and tool-making. In his workshop stood a line of glass jars whose protruding rims fitted into grooves of wooden blocks so that they could be slid off them by a touch of the hand. These jars were filled each with a different size and type of nail, while the jars, being open prevented the nails from rusting. This simple and completely effective contrivance, by which he also dispensed with the need for labels, was this young craftsman's own invention and gave me his measure more surely than did all the tools of his own making, his brass bowls, trailers and other devices. He was a mechanic with his roots in an uncorrupted craftsmanship and a craftsman who had adapted his manual skill to machines which, being small, had not compromised nor overwhelmed the human mastery and qualitative privilege. In other words, there was no break at all in continuity and the adze hanging up in the workshop was an anchor of security in the past.

I became more and more surprised at all I was seeing, and what I saw expressed a reconciliation between the old lost provincial life of the regional Surrey, crushed and all but extinct beneath the indifferent pressure of Suburbia, and the new life of our own period, but a new life which was a true evolution from the old without sacrificing any of its values or realities. The next thing I noticed was the extraordinary well-being of fruit trees – especially the open-air vines, including the Royal Muscadine and the mealy-leaved Black Cluster, which have the best-flavoured grapes and the figs fruiting in pots. I inspected, too, the flowering yuccas (*grandiflora* among others which flowers when it chooses, *filamentosa* which flowers at fixed intervals and *flaccida* which flowers every year) the alpines, ornamental conifers and tomatoes, the phloxes and hydrangeas, the lovely clematises, the peach and nectarine plantations, the fuschias and camellias, the herb garden, the vegetables (including the fine Mexican black dwarf bean, the yellow Chinese bean which ripens in the poorest of summers and is a very heavy cropper and the Mexican black haricot), and an innumerable host of other shrubs, trees, flowers and fruits at every stage upwards from the seedling in frame or glasshouse to the full-grown plant. The diversity was so bewildering that to attempt to cover it is the statistician's job but, thanks be, not mine.

Nor would such a list be at all comprehensive, since it would omit all those experiments in breeding new varieties by selection which are Mr. Jackman's own particular vocation and enthusiasm. He had, for instance, bred a magnificent rowan of dark olive-green leaves and dusky scarlet berries. It was selected from a mountain ash growing in the gardens – Sheerwater Seedling – and apart from the striking contrast of the bluish-green leaves and the brilliant berries, is a faster grower and with branches growing much more upright than the common rowan. He was also raising new varieties of the beautiful *Clematis campanile.* For staking he had discovered a special type of grass whose stems harden into a woody substance; he is experimenting with *Philadelphus grandiflorus* and planning to produce ash and hazel in his hedgerows and to put down plantations of sallow and basket willow for the same purpose. He was developing out-of-door viticulture and his empiricism ramifies into every direction and with all manner of plants. The gardens and plantations represented a happy blend between tradition and novelty, and I am accustomed to see them in bitter opposition, the one parricidally "liquidating" the other.

One of the sights of the Jackman Nurseries which particularly interested me was the condition of the soil, a kind of sandy loam. It was cultivated with a care, thoroughness and finish that produced a superfine tilth, and all the seedlings and saplings from the glasshouses and frames were planted out with enough space between each one to enable the rotary hoe and other small machines to pass comfortably between them and turn over the sod, the weeds with it before they seed. In an ironically paradoxical fashion the permit system, which, of course, has the effect of making the shortage of materials even shorter than it actually is, has had in the Jackman Nurseries the homoeopathic result of breeding its own cure.

The maker or grower, that is to say, has been forced upon his own resources and to improvise from the materials and conditions available at hand and at home. This improvisation has been the peculiar genius of Englishmen and particularly of the provincial Englishman which Mr. Jackman essentially is. Almost everything I saw in these nurseries in the way of equipment had been made on the spot and by the workers themselves. The very elm-posts of the great heather-thatched barn were from trees on the estate, the stakes being grown and the hand-made slats for shelter and fencing coming from a local firm.

The corollary to self-sufficiency in equipment of all kinds is, of course, the use of organics for the plants. When I saw such exuberance of growth in every type of plant, the slabs of rich colour in the fields, the vines and figs clustered with fruit, I began to be pretty certain that there was a slump in artificials in these Nurseries. As a matter of fact, none whatever are in use, whether in frames, the glass-houses or in the open. At one end of the gardens I saw an enormous compost heap, while leaf-mould from the coppices and woodlands in the neighbourhood and peat with sifted compost were the chief ingredients to feed the stock under glass. Mr. Jackman told me that the one gap in organisation was the lack of livestock and this he proposes to remedy as soon as may be instead of obtaining his organics from outside sources. He himself believes profoundly in the intensive use of our own soil and natural resources and (as all sane men believe) that it is sheer madness to cherish the illusion that we can depend indefinitely upon ill-fed foreigners to supply us with food and materials.

The Curious Traveller (1950:124-8)

Apples

CAN there be a more perfect symbol of late autumn than the apple? So I go to my apple-store like an acolyte to the shrine of Pomona and, if I do not wassail her, I mean to praise her on this autumn day. It took me a battle of five months with the bureaucracy to get a permit for my apple-house. Yet I might have done better to follow the method of Mr. Guy Speir of The Abbey, North Berwick. What he does is to place the apples stalk down on plain lawn grass and arranged in groups from each tree with the apples wide enough apart to leave air-space all round. He keeps the birds off by erecting four posts and hanging a net from them over the groups. Lord Cavan visited Mr. Speir and wrote to tell me that not only do these out-of-door apples withstand rains, frosts, snows and gales but greatly improve in quality. But what of the slugs and the field-mice?

Aligned between the rows of strawed slats in my apple-house are

Cox, Ribston, ribbed Gravenstein, brilliant Astrachan Red, bucolic Newton Wonder, the two Bramleys, including the Crimson subspecies, glowing Laxton's Superb, Red Russet, Golden Russet, scarlet John Standish, Egremont Russet, Allington Pippin, Ellison's Orange, D'Arcy Spice Apple or Baddow Pippin, Blenheim, Beauty of Bath, James Grieve and Worcester Pearmain, as goodly a diversity as, I believe, is to be seen in the homestead of any small grower in England. Most of them I grew myself.

The England of to-day is unworthy her apples because its economic system has victimised them. We waste millions of apples yearly for the good reason that our money-system regards the importation of inferior foreign apples as more profitable than the cultivation of our own. What is unquestionably true is that our home-grown apples ought to be the best in the world because our soils and climate are superior to any others in the world for the growing of apples, just as those of France are for the growing of pears.

Here is an example. I have in my orchard an apple that has become artificially rare in this country but was a common product of large gardens during the last century. This is Gravenstein, an oblong yellowish-green apple of delicate aroma and spicy flavour in the late autumn and imported (the trees, not the fruit) from Germany in the early nineteenth century. The original tree is said to have been growing in Germany about 1750. When cheapness, not quality, came to rule the economic roost, and English-grown fruit was left hanging on the trees to rot or lying on the ground to be eaten by slugs and mice, this species, abandoned in despair by English orchardists, was transplanted to Nova Scotia and the southern part of U.S.A. The fruit is now or was up to recently imported thence in large quantities. Had I not grown the apple myself, I could not have known the difference between the English and the imported Gravenstein. In the latter, this aristocratic flavour is lost and the fruit is virtually tasteless. Yet the Gravenstein is very easy to grow, is not subject to scab or canker and in my experience has only one fault – the tree is rather butter-fingered in holding on to its fruit. But this matters very little with so prolific a kind.

That is one of the disastrous results of the cheap food imports policy; it has destroyed the discriminating faculty of the English palate. With the possible exception of Delicious from North America I would not barter a single good and sound English apple for fifty foreigners. No doubt, apples grown and eaten in their native countries are palatable enough – American Mother, for instance, has a good reputation in the United States. But importation ruins their flavour, partly but not only because they are for keeping purposes so often injected with chemicals. Even so, I do not believe that any apple in the world, though not imported, can compare with the finer English ones. There is a richness, a special and penetrating quality about an English apple, picked, preserved and eaten at the right time, that no foreign country can rival. But because we have thought so long about food in terms of ships and shops, we no longer know the difference between good food and bad.

The English apple, too, has a history and a tradition unapproached by those of any other country. Before it was ousted by the German Christmas Tree, our native Kissing Bough was hung with candle-lit apples. We had Pearmains and Costards as early as the thirteenth century, the latter used for pies and giving "costermonger" to the language. By 1500, Pippins were being grown from seed (hence their name) and breeding true, the Golden Pippin or Great Golding being the praise of the orchardists and herbalists for centuries.

In Elizabethan times, the Pomewater, the Queening, the Codling, the Leathercoat and the Apple-john had made their way into literature and the last three into Shakespeare ("I am withered like an old Apple-john," says Falstaff), while Codlings gave us "codlins-and-cream" (because of its aroma) for the willow-herb. I have a Codling myself in the shade of which my favourite sheepdog is buried. It is a capricious tree, bearing heavily or not at all as the whim rather than the season decides.

But the seventeenth and eighteenth centuries were the great ages of apple-planting and experimenting with new species and varieties because in them the country manor reached the climax of its long and illustrious flowering. Nonpareils, Russets, Kirston Pippins, Stokens, Cockagees, Hollands, Quarrendons (still grown in the West Country) and many other varieties made their appearance. Lord Scudamore at Holme Lacy in Herefordshire greatly improved the native cider apple by importing specialities grown in Normandy and crossing them with the home stock. The famous Redstreak cider apple, praised by John Philips in his poem *Cider,* was the creation of this Herefordshire squire. The empiricism of John Evelyn was responsible for Pomme d'Api or the Lady Apple. According to Merlet, the Api was first discovered as a wilding in the Breton Forest of Api. Dr. Hogg mentions the claim that it was brought to Rome from the Peloponese by Appius Claudius. It is a beautiful little dessert apple in season from October to April. The Ribston Pippin was first grown at Ribston Hall in Yorkshire in 1707 and bred true there for 220 years. I believe that one of the original trees still bears there. But because these plantations of the Carolean, Queen Anne and Georgian squires were not commercialised, experiment in new varieties had free range and specialisation in a few types for their market value had not yet ousted quality for quantity. The dealer had not yet become a more important person than the producer.

Sowing pips, selecting scions, grafting, crossing and hybridising continued in the nineteenth century until the collapse of agriculture in the eighties. This policy destroyed the culture and economy of the countryside that had gathered strength and enriched the whole nation for a thousand years, and apple-culture suffered with them. It was before 1879 (when the first crash came) that Pomona's kingdom flourished and extended her territory. In 1800, a genius among pomologists, Thomas Andrew Knight, succeeded in breeding a number of other new varieties. One of these, Yellow Ingestre, has precariously survived and is still grown in the private garden of a friend of mine. It is a delightful little yellow apple, very sweet and crisp in September. Our familiar Cox's Orange Pippin was first bred in 1830, the equally celebrated Bramley's Seedling, still our best cooker, in 1876, Worcester Pearmain in 1874 and another first-class cooker and long-keeper, the sturdy Annie Elizabeth, in 1864. Blenheim Orange was much earlier than any of these, having been raised from a pip at Woodstock in 1818, and the noble Claygate Pearmain was found growing in a hedge in 1823. From the inspiration of Knight, Charles Ross, born in 1825, raised seven more varieties by hybridisation, one of which, a cross-pollinator of Cox, bears his name. Another of the great names in pomology is, of course, the Laxton family. Thomas was born in 1830 and his descendants were responsible for such well-known and excellent apples as Laxton's Superb, Epicure and Lord Lambourne, the majority of them obtained by crossing Cox with a number of other pippins.

It is interesting that the great orchard experiment brilliantly described by Mr. Hugh Quigley in *New Forest Orchard* owed its success in the main to Miller's Seedling (raised as early as 1848) and these three Laxtons, in spite of the advice of the experts against them. Crabs really deserve more than a passing mention. Experiments with them from the Middle Ages onwards culminated in an enormous variety of cider apples for making verjuice as well as cider. One of these, Tom Putt, still widely diffused in the West Country, is as good a cooker as cider apple. Most of these cider apples, the majority of which are as obsolete as the art of raising them, possessed rich country names, some of them preserved in Hardy's novels. A host of skilful craftsmen, as forgotten as the rural culture that was their seed-bed, co-operated with English soil in making our native Pomona the queen of the world's apples. She appears in all her glory in Samuel Palmer's picture, *The Magic Apple Tree.*

The few modern varieties, of course, come not from the country house but the research station, for instance, Taunton Cross and Worcester Cross from Cox, Early Worcester and Late Cox, varying in season rather than in flavour and type from their originals. The two most promising lines of research in our day are to breed frost-resisting varieties and to elaborate dwarfing stock like the Dwarf Pyramid No. 9 of East Malling. Cordon and espalier apples are, of course, an enormous boon to the small grower, not only by economising space but dispensing with the ladder in pruning and gathering. As among other fruit crops, disease has increased beyond measure and a huge industry is devoted to turning out sulphurs, tar-oil distillates, nicotine washes, lead arsenate sprays, D.D.T. and other poisons. Spraying once a year has now become seven times a year. The organic orchardist is caught in a dilemma here. He realises the anti-natural bias of inorganic science which seeks to defeat rather than to understand nature, that prevention by health has not been its aim and that these poisons are used at the expense of the "ploughs of God," the greatest of all humus-makers, the earthworms. Nor can a biggish orchardist lay down waterproof sheets at the foot of each tree to prevent the soil from being contaminated by insecticides. On the other hand, disease is so rife in the modern orchard that the trees, however robust, are constantly exposed to infection from orchards that go the whole modern hog with chemicals. I myself only use harmless but effective sprays like Voelck, good for aphis and even red spider if used before they get their claws in. The trouble is that modern cultivation has become so thoroughly artificial that incidence to disease has got into the *stocks.* Few are immune and natural resistance to scab and canker (as Mr.

Quigley ruefully remarks in *New Forest Orchard*) has not received the attention it should have done.

As H. V. Taylor has pointed out in *The Apples of England,* apples are sweet or sharp, acid or sub-acid, delicately or strongly aromatic, while some cookers froth on the plate and others like Tom Putt, Warner's King and Annie Elizabeth, are ideal for dumplings. Every farmer, says a modern biologist, should be a bit of a poet and surely every orchardist ought to be something of a connoisseur in flavours to round off the anxious joys of growing apple trees. There are some apples that taste like nuts (Blenheim and the walnut-flavoured Orleans Reinette, for instance), others like peaches (Melba), others again like strawberries (a Herefordshire Worcester Pearmain eaten at exactly the right time), and one like hock (Gravenstein). But apples have arts of pleasing the palate that go far beyond such classifications. The really creditable orchard should always have a few Blenheims, Laxton's Superbs and Ribston Pippins. The bouquet of a Blenheim we all know, but the January Superb has a subtle aroma in the mouth (not to mention what Sir Thomas Browne called "the handsomenesse of the same") like, but more refined than, that of Cox. I find it, too, a very hardy tree that clings most tenaciously to its lovely fruits in the bluster of full gales. My small Superb had a crop of eighty apples this year and lost only four of these in frantic wind-storms. As for the Ribston, it can be a grand cropper so long as it has early pollinators (Lord Derby, Grenadier and one other I forget) and is very little if at all inferior to Cox if eaten before it surrenders its juices, as it is liable to do at the turn of the New Year. Here are a few other master-apples that every fastidious garden should grow – Golden Pippin, King's Acre Pippin, Golden Reinette, Cornish Gilliflower, Laxton's Epicure, King of the Pippins and Gravenstein.

A modern loss is the neglect of the Russets. I suppose one reason is that in the way of looks they have little to recommend them. They have none of the meretricious glamour of Worcester Pearmain, which nine times out of ten is a worthless apple. Like the plane tree when it has shed its bark, the Russets are inclined to have blotched, spotty and leprous skins. But they are superb eaters, sharp-and-sweet, rich in juices, spicy, invigorating. The best of them known to me is actually a very handsome apple and in my view superior to Cox, and Cox is a capricious apple both for taste and in cropping faculty. This is Orleans Reinette, a Russet that goes back to 1776 and probably further and is the colour of old gold washed over with russet. I had four of these exquisite apples sent me from a private garden in Norfolk, and it gives me a pang that I shall probably never taste another, since the privileged garden that grew them is to be sold.

But there are plenty of other Russets that would make the modern table more of a grace – Wyken Pippin, Nonpareil, Sturmer Pippin, Rosemary Russet, Aromatic and Golden Russeting, Egremont Russet, Court Pendu Plat and D'Arcy Spice or Baddow Pippin which, since it came from the Baddows in Essex before, I think, being grown by the Eves of Tolleshunt D'Arcy, is its proper name. Why are these apples not the pride of many an orchard? *Herefordshire Pomona* says of Golden Russeting that "it hath no compare." Russets are not at all troublesome as croppers and most, especially the Baddow Pippin and Egremont, are worthy keepers, lasting well into spring. They are more resistant to frost than Cox, Bramley, Allington, Bismarck, Warner's King and other modern apples with an incomparably wider vogue. D'Arcy spice is as good an apple in May as in January. Court Pendu Plat has the advantage over all other apples that it is the latest of all to blossom, later even than Crawley Beauty, and so entirely escapes the late spring frosts. For its discretion it is known as the Wise Apple.

D'Arcy Spice Apple raises another issue. It has a lyrical flavour but takes kindly only to the soil of Essex, being a parsimonious cropper and biennial at that elsewhere. What I should dearly like to see revived is the sense of the regional apple, a sense that nowadays only survives in the West Country. Though in sparse numbers, the regional apple still exists – Baddow Pippin from Essex, Maid of Kent, Maidstone Favourite and Sunset from Kent, Proctor's Seedling from Lancashire, Cornish Gilliflower and Cornish Aromatic, Tyler's Kernel from Herefordshire, Colonel Vaughan from Sussex and Tom Putt and Sops-in-Wine from Somerset. The fact is that apples are never really at their best unless they are grown in their own local soils, some, of course, less exactingly so than others. The analogy with the great wines of regional France is exact. When England is forced by circumstance to look once more to its own land to support it, as is bound to happen in the near future, then I hope we shall set to work rediscovering in what regions apple varieties best like to grow and replanting them there. That assuredly will be one way of reducing disease. It will even be worth growing Worcester Pearmain when we confine it to the Old Red Sandstone, and pick it two months later than the books instruct.

There is a host of pomological problems for a wiser posterity to investigate. What soils, for instance, do apples prefer as a class? I find that, if mine get their feet into clay, they develop canker and are best cut down and cast into the oven. On the other hand, Mr. Hugh Quigley has found the reverse to be true in Hampshire. This kind of problem is far more important than that of the influence of chromosomes in transmitting hereditary characters or what varieties stand up best to being standardised and mass-produced. And we shall be making our way towards sanity again when we take to heart the salutary truism uttered by a great plant-breeder, Bunyard of Maidstone: –

> "No fruit is more to our English taste than the Apple. Let the Frenchman have his Pear, the Italian his Fig, the Jamaican may retain his farinaceous Banana and the Malay his Durian, but for us the Apple."

An Englishman's Year (1948:28-35)

Ecology

IN my county, I know personally but one other man who has the same interest as I have in what are called bygones, or makes use of them as I do, namely by bringing them up to date. He is the agent or manager of the Verney Estate, and he has furnished his whole cottage with them. Everything in his house has been procured by looking for it himself, mostly at farm-sales in Lincolnshire. When I visit him, I see fine 17th and 18th century yew armchairs, a Regency chest-of-drawers of fruit-wood with bronze instead of brass handles and poker-work between them and on the cornice for which he paid 7/6, an Empire sofa for which he paid 2/6, a great elm-bowl in which he keeps his quinces and not dissimilar from my oaken one, apple bowls, Worcester china, Windsor chairs for which he paid 1/6 the set and recaned himself. Every stick of furniture, his bedspread, the pictures on the wall, all are bygones. The link between us is yet closer, for his carved butter-stamp, a vertical wheel ornamented with corn-sheaves and attached to a two-pronged wooden handle, is now one of *my* bygones. The very clothes on his back are home-made. The linen was woven from his own flax and the tweed from his own sheep. The overcoat he had worn for thirty years looked as good as new. He made one reflect that the modern system of supplying massed populations with cheap devitalised foods, cheap synthetic medicines and cheap clothing as ephemeral as the houses they live in is more expensive and impractical than any the world has ever seen.

The relation of this man to his environment was the most ecological (ecology is the organic relation of living things to their environment) I had ever encountered. His lath-and-plaster cottage in a village mentioned in Domesday Book was as perfectly adapted to its natural setting as the furniture to the house and he, the apex of the pyramid, to all. Nowhere was there an alien note. This wholeness was carried through into the very meals he ate and thoughts he spoke. I linger in memory over those meals I have partaken with him. For teas we sometimes had sugar-beet-chutney sandwiches with compost-grown celery. Every mouthful had a double value, its own and that of the beautiful stuffs and ware, integral with the fields outside the window, round which and from which we ate it. For lunch we would always eat off the estate, straight from the flat Swedish cooking stove. I became acquainted with some of the varieties of cottage dishes from home-bred pig with home-grown vegetables as described by Walter Rose in *Good Neighbours*. In summer we would finish a large bowl of strawberries inundated with cream from the Guernseys; in winter, a quince tart, the pastry made from the home-cured lard of the home-pig or from a mixture of home-cultivated wheat, oats and barley. The quinces from the cottage garden in chunks of a mahogany brown tasted like the finest preserves. What with the cream and all, I would after these princely-peasant meals gently resist his intention of taking me a tour of the estate. We have got out of the way of taking cream. When it is poured out on one's plate out of a full pint-pot (with more to follow), the afternoon should be devoted to rest.

The health and variety of the fruit and vegetables growing in his garden and the feats accomplished with cuttings were a constant wonder to me. It all seemed the work of a magician. In summer, it was as clean as a new pin; in autumn, covered with weeds – for green manuring. The cuttings were of every variety of fruit-tree, bush and shrub, the fruit-bushes bearing when only some inches high. Apples clustered like onions round a stick on apple-saplings without any lateral branches. He told me he always used to make these new gardens out of old or derelict ones wherever he had lived and usually left them to take up a new job just when they began to reward his loving husbandry. But he never minded about that. After all, he said, look what our forefathers have left us!

After one of these meals, we would, instead of seeing self-sufficiency in action outside the house, discuss it inside. The first time I took one of these rare traditional meals off his kitchen table, I remember noticing a hand-mill with a great wheel screwed to another table. For this agent is not to be caught napping by the gods of big business. Let them get on with their games, is his idea, but let me get on with the job of running the estate as it should be run. Yes, said I, "but what about when they interfere with you?" "That's why I fixed up the hand-mill," he said. My host is a very simple man – that is the point about him. He is not impressed with a thing because it is big, and covers vast distances. Nor because its dividend percentages run into two or three figures. He is so simple that he believes an ounce of quality to more than worth a pound of quantity.

Sitting round his log-fire in winter or in his cottage garden in summer, we would talk business. After the hand-mill, we naturally began with bread, that burning question, what Isaiah called "the stay and the staff, the whole stay of bread," but what nowadays might be called a broken reed. My host had to sell the wheat off the estate at 14/- a cwt. It then goes off to the big roller mills at Hull, Cardiff, Liverpool, or other "wen." By extremely devious ways (devious not only in the sense of travel but what is done with it), it returns where it came from as flour and is bought back as flour at 2/2 per 7lbs. Possibly, if bought by the cwt., it might be got at 28/-, that is to say, at exactly double the price it was sold. Now if this corn off the estate had been ground by a mill *on* the estate or in the neighbourhood, it would have cost 16/6, 14/- for the corn and 2/6 for grinding, while the whole of the costs of transport would have been saved.

But that is by no means the whole of the story. The corn returns as flour not only at double the price but half the value. The simple reason is that, between leaving and returning, it has been deprived of the wheat-germ, containing ten vitamins. In war-time it has had

25% foreign wheat added to it (so that the risk and expense of ocean transport has to be super-added); it has had milk powder mixed with it and calcium naturally present in vegetables artificially added to it. Its weight has been increased by something like 10% of water. Lastly, the foreign wheat in it is reaped out of exploited soils whose low yield has been yearly forced by heavy druggings of artificial manures. Yet this shadow of the substance of the bread of life is a better thing than pre-war bread, even though the flour goes bad unless baked at once. This was nothing more than chemically bleached starch or, as it has been called, "blotting-paper bread," and was, probably enough, one of the causes for the decline of the birth-rate.

Not only is this peculiar system called economics but it is even regarded as "inevitable." It is part of "the march of progress," as though the stars had ordained that modern Englishmen should eat bread with the life taken out of it or the wheat from the fields was lost in the bread on the plate by the workings of natural law. Moreover, those who work out these simple sums are regarded as romantics, cranks, laudators of past times and interferers with the clock. They are even called the dangerous fomenters of "economic aggression." But perhaps these derogatory terms are applied because, if this kind of simple arithmetic were to become at all prevalent, it would interfere with something else beside clocks. Twenty six thousand country mills have not been put out of action for nothing . . .

How had the rot set in? Nothing but an examination of the whole modern system of economic parasitism can reveal that. But I communicated to my host what I had gleaned of the actual process of disintegration from a letter about a region in Ireland. There was a water-mill every three miles in which wheat and barley were ground and oats crushed. The loaves were baked in the farm-kitchens, some of the mills removing the coarser bran. Then the millers began making white flour in the bigger towns. It was less trouble to buy a sack of flour than to cart the corn to the mill. So the farmers ceased to grow wheat, since the millers only bought local grain when the price was low. The bakers in the towns sent white loaves into the country and the farmers bought them now and again, then more, then exclusively. As no bread-baking was done, no buttermilk was needed; all the milk went to the creamery and household butter was bought. As the creameries work only in summer, the farmers adopted summer dairying and grew fewer oats and roots for winter keep. Less tillage meant fewer horses. The journey to the creamery was for bringing imported maize to feed stock, while the barley, no longer ground, was sold to the brewer. The brewers' demands fell and less barley was grown. Meanwhile, the country mills had ceased to function and fell to ruin. When bread and butter-making ceased, the household milking was done by a hired worker and the girls in the family went to jobs in the town. Only Mary was left and when it did not pay to keep her only to look after the chickens, Mary went and then the chickens. At the beginning of the war, farming there was keeping a number of cows for summer milking and raising young stock on grass. The population that should have been producing food and consuming farm products had emigrated. This was what I meant when I said . . . that, when the country millers were forced out of business by the combine underselling them, the linch-pin fell out of the rural structure. It can be restored by cultivating the principles underlying and overlying the cultivation of a garden and by no other way.

We gourmets had many such post-prandial talks. Gourmets we are, since made-up delicacies with high sauces and fillips to the palate are but the decadence of food. When you eat real food, you are taking in not only the food but the landscape out of the window and the sun that ripened it and the rain that invigorated it and the air it living breathed and the husbandry that fostered it and the creative power of the earth that developed its vitality. My host had lived up all his life to his ideas of going straight to the heart of things, more than I have ever been able to do. Consequently, there was something about him which made you believe in him. A strange, long-faced, bearded man of austerity and realism, a man who followed what he called the "golden rule." Nor do I idealise him; he is the French peasant-leader rather than the English, and once farmed his strips in a French commune. Perhaps that is why his attitude to birds is repugnant to me; it overcomes his profoundly religious sense and short-circuits, I think, the comprehensiveness of his husbandry. But as the gardener who farms or the farmer who gardens, he is a prince among men, the prince of the peasants. Since the future of England depends not upon her industrial superstructure which is "the primrose path to the everlasting bonfire," but on her agricultural foundations, he is a man of consequence.

This Plot of Earth (1944:226-31)

The Country Workman

WE stood by the farm-gate, the small farmer, the carpenter-builder and I, and talked about the village crafts. What killed them, said the farmer, was when the son, educated at a school away from home, deserted his father's trade. When he was brought up on the farm or in the workshop, that is to say, at home, he absorbed his father's trade soon after his mother's milk. His father would go away on a job and put his son in charge, or his son would do the odd jobs until he was ready to step into his father's shoes. This simple process was the most effective method of linking up family with livelihood, home with work. It preserved the one through its interaction with the other. It was thus a continued guarantee, through the interdependence of the crafts with agriculture as the common focus of them all, of the stability and vitality of the organic village or small township. It was, too, a vocational training which kept the balance between hand and brain and fused character and responsibility with skill. Though it conferred the freedom of expressing "creative energy," it was a freedom controlled by the needs of the local community and the discipline of the home.

To this unity between creation and continuity the modern scene has been consistently hostile. Consequently, industrialism has not contented the worker, be his wages high or low, and the drift is right away from personal responsibility. The relationship between father and son, master and apprentice, home and trade, is, by preserving that responsibility, part of the timeless order of society. In terms of former piety, it was ordained by God.

Every country workman who is something more than a mechanic has two primary endowments, an organic relation with the earth and an hereditary proficiency. I take as my first example a jobbing gardener of very limited intelligence. He is a man incapable of abstract thought and would make an indifferent witness in a court of law. When he has any information to give, he usually repeats it four times, not only to make quite sure that what he has to say is not missed but that he himself has got it right. He would take an hour to read the page of a book. To give cogent reasons for the right principles of husbandry he holds would be as much beyond him as for him who fetched the worm for Cleopatra to grasp why she wanted it. Yet his instinctive familiarity with the laws of nature is profound.

I received recently a load of mixed timber for firewood which this man cut up with and for me. It consisted of unstripped poles, 12-16 feet long, ready for sawing into logs. All these he distinguished and name at once. Not only so but whether the trees from which the timber came had been saplings or full-grown and whether before felling they had grown in sunshine or shade. The maple, for instance, has deeper corrugations and is paler in tone if the sun has struck upon it. He could identify one species of log from another simply by striking the two together, the harder woods like whitethorn having a crisper and more percussive note than the softer. Barks like whitethorn and field maple, that in the pole look indistinguishable, he knew apart by the more flakiness of the former. Ash is easier because it nearly always betrays itself by the pittings of the burrowing beetle, while even a novice in wood-lore could hardly fail to recognise a whitethorn pole from its flattened and slightly fluted bole. In time he would gather that the silvery wash over a corrugated pole, not unlike the tone of Kentish old dressed oak, revealed a sapling of hedge-elm.

But this man knew a rotten log which lacks this silveriness to be elm, not by the bark but the cup-shaped markings where the wood had broken off. Concentric red lines in the flesh of a sawn portion were a sign to him of damson or sloe. But though the lines of both are red and semi-circular, he never confused sloe with damson. Damson, again, resembles buckthorn in possessing these concentric rings, but buckthorn only on one-half of the bole, the half that has faced the sun. Though buckthorn is not a common hedgerow tree in my neighbourhood, he never made a mistake about it in the pole. Other woods he would know by stickiness or sponginess of their resistance to the saw, very different from the whitethorn that offers a smooth and clean cut. Others, again, he would know by the mosaic of the bark corrugations, others by their habit of growth, undiscerned even by the naturalist or the forester who would understand the growing tree but not the pole. He could even tell the age of a tree from which the pole had been chopped by the differences between the sides and centre of a sawn portion. What is more wonderful, he was never beaten by logs or poles even when their bark had been stripped, though he had to think twice when they were beech or ash.

Knowledge of this kind, which verges on the mysterious, cannot be acquired by learning to identify one tree from another, whether standing or felled. It is derived from a certain inborn sympathy, strengthened by habitual observation, with the very pulse of natural growth itself. It divines nature's own indwelling pattern of life which, like the human pattern distinct from but allied to it, reflects the Natural Law.

The Wisdom of the Fields (1945:52-4)

Boredom

I have yet to meet a bored countryman. Worried, yes, anxious, continually persecuted by ignorant officials, overworked, even despairing, bitterly cynical about the future, even resolved to leave the land because of bad conditions if he is a labourer, of insecurity if he is a farmer, of crushing taxation if he is a good squire. But boredom is not a word that can be found in the true countryman's dictionary. Why? Because except on the most highly mechanized farms all countrymen follow ill or well Voltaire's maxim of cultivating their gardens. They take, that is to say, pleasure in and exert skill on their jobs. But when, say, 90 per cent of modern workers have been deprived by the machine of skill and interest and pleasure in their daily work, they are suffering from the disgraceful sin of being bored. This explains at once why, when they are released from boredom of their daily work, they depend upon the mechanical hedonism of being amused by others. But it does not release them from the disgraceful sin of being bored. This is the nemesis of modern urban civilization which has all but killed rural England: it is delivered over to the disgraceful sin of being bored.

This Plot of Earth (1944:248-9)

The Monotonous Landscape

MANY judgments have been levelled at modern industrial civilisation. Not the least of them is its monotony. What varieties of building survive in modern cities are none of their making, and the more modern they are the less distinguishable they become. The suburb, the building estate, the factory, the cinema, the government office, the department store, the aerodrome, the railway and wireless stations, these have no frontiers. It seems only by accident that their occupants speak different languages. They are the repetitions of Cosmopolis. Manchester might as well be Montreal, Stalingrad Sunderland. Identity is meaningless; there are only distances to and from the same place. Different places are interested in and so inclined to like one another. Those who live in places separated by miles, not character, are inclined to shed their human differences in the sense that they do the same things in the same way. They are populations rather than persons, and they do nothing. They do what their industrial economy tells them to do. What they are told to do has nothing to do with what they would naturally choose to do, and in past ages could do, and loved to do by virtue of being human beings. Doing the same things day after day, they are bored – bored, as the saying is, to death; and death plays a very large part in modern civilisation. It is mainly concerned with inorganic quantities. These are predictable because they always operate in the same way. So do the people who manipulate them. So hate abounds.

This boredom and this monotony are being steadily diffused into organic nature, the original home of interest and variety. So the incidentals of work – wages, costs, output, quantities – become, as in the town, paramount. Its essentials – what kind of work and how it is done – disappear. To measure human labour in terms of horse-power-mindedness and the bulldozer standard is merely to magnify monotony. A similar transformation affects the face of the country as it has done the various appearances of the old towns. If there is a wood, they are all the same trees; if a meadow, it is composed of a very few grasses instead of many, and many meadows are merged into one field. The cities not only spread their own sameness over the countryside and suck the rural diversity up into them, but what is still country becomes the same country. Hedges, those manifestations of difference, vanish. Lanes as tortuous as rivers are straightened out. Wild animal life becomes restricted to a few species like the few grasses in the fields. Utility supplants use; profit pleasure; expediency a way of life; efficiency, which is cost-cutting, craftsmanship. Such are the triumphs of dullness and sameness. Modern novelties like the bulldozer, the multiple plough and the combine harvester are like headlines that make the text describing a variety of doings superfluous.

Where Man Belongs (1946:115-16)

The Curse of the Conifer

SUDDENLY, all was changed, as though a traveller through this wild Elysium had reached the edge of Hades. What I saw so sickened me that I was glad after some miles to turn back before reaching the open mountain moorland.

As I ascended the Valley, I did see a few sheep, but the wonder was how even these few could possibly pick up a living. Turn after turn of the winding defile cut by the beautiful stream leaping down from its source on Rhiw Cwmstab was clothed or rather smothered like Desdemona by her pillow in larch and spruce, and so densely that all variation such as every mountain valley gives had been blotted out. The original landscape had been effaced and what had taken its place were walls of light green and dark green in vast uniform blocks on either side of the river. Where the scene was more open to the east, the hillside was strewn with the fallen logs of former oakwood, and sometimes this derelict terrain had been planted up with seedling spruces with the logs left to lie all about them. To any lover of natural beauty the valley would have appeared as transformed from a place of freedom to a prison. It was smothered; it could no longer breathe.

The spectacle of this murdered valley is no isolated example. If sheep-farming has been all but obliterated here, what will happen to the wool and mutton of Wales when 800,000 acres of it are planted up with conifers in accordance with the policy outlined by a Forestry Commission official who has declared: "We intend to plant 800,000 acres in Wales. We intend to change the face of Wales. We know there will be opposition but we intend to force this thing through"? And what will the face of Wales be like for holiday-maker, tourist, native, farmer and indeed for any human being who is not "a dull finite clod, untroubled by a spark", when this martinet policy is carried out and for mile after mile and as far as the eye can range, everything visible will be a monstrous uniformity of the light green of larch and the dark green of spruce? Is it for this that we propose to squander millions of money, to sacrifice good mutton, lamb and clothing, to blot out nearly a million acres of a country renowned all over the world for the supremacy of its beauty? Are pit-props our only form of wealth? When it is considered to what uses we put our land through the ages, such wealth is poverty indeed. Correlate the dispossession of sheep-farmers and the felling of oakwoods (Great Wood in Sussex, for instance) to plant conifers with the artificial deserts of opencast mining and ironstone mining among the wheatfields of Northamptonshire, the hydro-electric schemes of Snowdonia and Plynlimon drying up the rivers, the absorption of yet more agricultural land for new towns and industrial purposes and it becomes clear that we are waging a merciless war upon our own country, fouling our own nest, in pursuit of the phantasm of export markets and of an industrialism expanding against the reason and nature, against out heritage and the means to live. Out of the fields that feed the people we create sterility and out of the wildness of natural beauty a wilderness.

You have only to look at the floor of a spruce plantation to see that the fir needles choke all growth and become so acidulated that the land when the trees are felled in quantity (as they must be for industrial purposes) will be unable to resist erosion from wind and rain. These needles are a Nessus shirt to the skin of the earth. It is calculated that in a generation the human population of the globe will be increased by a further 800,000,000, each one with a mouth and a belly. At the same time, nearly all nations and peoples are industrialising themselves as fast as they can make machines and build factories. How are these multitudes to be fed when more and more of their food-bearing lands are offered up as a sacrifice to the mono-idea of industrial expansion? And how can these myriads export food to us when they themselves have not enough to eat? In the face of such ineluctable facts, the war against our own fields can only be defined as a monomania. It is not a policy but a possession. It compels us as in a dream to squander fortunes in tearing out the fertility of Africa and to lay our own fruitful land waste. Centuries ago, Petrarch wrote to a friend, "Would that you could know with what joy I wander free and alone among the mountains!" But that was while the bride of quietness was still unravished.

I wrote this personal account of the Forestry Commission's particular policy in the Grwyne Fawr Valley and general policy in Britain before I had seen Dr W. H. Pearsall's *Mountains and Moorlands,* before indeed it had been published in 1950. Since Dr Pearsall is a Doctor of Science, a Fellow of the Royal Society and Quain Professor of Botony at London University, what he has to say about the biological effects of the Commission's conifer plantations is of great importance. I summarise his conclusions.

(1) The policy is purely commercial and uniform conifers are planted because they yield a crop in the shortest possible time.

(2) Since the trees are grown in homogeneous blocks, they are clear-felled at maturity and form a "close canopy" which has "the great biological disadvantage" of suppressing all growth on the forest floor and so "any natural regeneration".

(3) Since there is no natural regeneration in such woodlands, there is no maintenance of woodland humus which prevents the soil erosion and the leaching "which follow the exposure of the ground on clear-felling".

(4) The absence of ground-flora, partly from the dense shade and partly from the carpet of conifer needles, gives "an acid surface layer" and tends to form *mor. Mor* soil is highly acid, lacks earthworms and nitrates and contains only a few bacteria. In other words, there is no defence against soil erosion and soil erosion on 800,000 acres in Wales alone is serious enough to justify the utmost opposition to the Commission's policy, quite apart from the destruction of sheep-farms and the desecration of landscape it entails. Thus Dr Pearsall declares in scientific terms what I have done in my own.

It is evident, therefore, that afforestation by means of uniform blocks of coniferous trees has no biological justification, especially when such blocks are planted on slopes which, on clear-felling, invite the speedier erosion. In his book, Dr Pearsall draws special attention to the general ecological degeneration of mountain country, peat and bog conditions everywhere displacing woodland and grass, a few species displacing a variety of them, lowered nutritive levels displacing former ones of a higher fertility and acidity a more balanced humic content. Human agency has played by far the larger part in this deterioration. In the light of his investigations, the conifer policy has to be viewed not as an example of man's progressive mastery of his natural environment, as is the accepted opinion, but as hastening the diffusion of those desert conditions which have been so marked a feature of industrial civilisation. This is a process that can have but one end and is initially due to the dominance of the economic motive over all others, of the hypertrophied and, one might say, cancerous growth of one element in the complex human make-up to the exclusion of those cells of human activity which are not economic at all but have a vitally important part to play in the well-being of the human and social organism. The fundamental danger of the universal extension of State control is that it expresses the final stage in the exclusive sovereignty of Economic Man, final because it carries the exploitation of the earth to an extreme far beyond the safety limit. Economic man in starving out the other more qualitative elements of man's being starves himself. The conservation of natural resources is not only a counsel of necessity but a means of restoring to man's unbalanced spirit those qualities of health and sanity by which he may become himself again.

The Southern Marches (1951:126-9)

Return to Husbandry

OUR agricultural self-sufficiency was destroyed by the acquisitive economics of farming out foreign soils to feed or rather mis-feed the mass-proletariat created both by the Enclosures and the Industrial Revolution. A usurious system was built up round this primary sin of abandoning our native land. It not only maintained itself by ruining our own farmers and pushing those of other lands into debt, but handed over all the power and credit in the community from the primary producer to the dealer. Examine the vested interests of this country, and it will be seen that they are nearly all clustered round the breeding of money, transport, distribution, and other secondary activities to the depression of the creative elements in a nation and the neglect of real needs. There are many reasons why this system has come to the end of an overlong tether – soil erosion abroad, liquidation of foreign investments, the building up of manufacturing industries in the countries that exported cheap food to us either to pay interest or exchange it with our industrial goods, and other phenomena. There are only two alternatives to it. One is that the State shall organize the whole country into a kind of gigantic exporting firm ('export or die'). It will hunt the world for markets which no longer exist and depend on an international money-shop, conscript fluid labour forces, dole out 'social services' as a palliative for serfdom, and finally crash or become embroiled in a new world war which will be the epitaph of all civilized communities. Such a policy only means that the State and the combine take the place of the *laissez-faire* individualist, with the cushion of 'social security' to soften the transition. New Presbyter is Old Priest writ large in a Blue Book for gospel and a State Temple with a congregation of the money-changers for a church. The other alternative is a return to husbandry.

We gather, then, that 'husbandry' is a larger term than one equated to the restoration of a self-supporting England, based on its own fertile soils and receiving what imports we need by the exchange of surpluses on the lease-lend principle. The proper balance of town and country, the full development of the home market, agriculture the *only* primary industry, the abandonment of the idiocy of long-distance farming by urban clerks and officials who try to cheat nature with their own little industrial gadgets, the recovery by the country of its indispensable self-government and thereby the recovery of local and personal responsibility, these are all contained in the term, husbandry. But its full meaning is not coextensive with them, and has a wider circumference. The social creditors who propose to rescue the national credit from private monopoly and the debt-system by the issue of notes to balance production with consumption are preparing one way for a return to husbandry, just as self-sufficiency is preparing another. But these measures are no more than preliminaries; all we can say is that the comprehensive practice of husbandry is impossible without them. Modern power-farming to meet the shortage of imports from abroad has little or nothing to do with husbandry, mostly nothing. It is merely an application to nature and the soil of urban and industrial methods of converting inorganic raw material into goods.

We shall not, in fact, begin to understand the meaning of husbandry unless we relate it to the first principles of the natural law, which is an earthly manifestation of the eternal law. This closely involves a study of natural processes by biological tests from which orthodox science, as distinguished from the newer science of certain pioneers whose guiding principle is the rule of return, has widely departed. Therefore, the meaning of husbandry is a fundamentally organic one, and neither chemical nor mechanical except in so far as chemistry and mechanics serve the living organism of nature and the soil. At present, their object is the forlorn one of mastering it, and this by overriding the natural law is doomed to failure by soil-exhaustion. Again, this meaning is 'ecological,' which means the relation of an organism to a particular locality which favours its due expression. The pattern of life worked out by pre-industrial rural society was an unconscious obedience to ecological laws because the interdependent nuclei of the pattern as a whole were localized. A centralized agriculture except for purposes of supervision is a contradiction in terms, as well as of the natural law.

If we look well into the word 'husbandry,' we can risk a definition of it, namely loving management. It means man the head of nature, but acting towards nature in a family spirit. Nothing could be further from its meaning than the modern and scientific 'conquest of nature,' which is not only contrary to the natural law, but an absurdity. Modern secularism debases man by making him purely the creature of earth with no destiny beyond it. At the same time, it elevates this reduced animal beyond his station by making him the conqueror of nature – an altogether childish conception. These illusions of thought come from trying to break through first principles. But loving management exactly defines man's place in nature, and so honours the natural law, which regards man as chief of the creatures of earth, but subject like them to their Creator.

Such management includes personal and local responsibility, or it means nothing at all. The most effective way of ensuring that responsibility is by ownership, or, as second best, by security of tenure. Here, too, we have guidance and precedent. The old rural society, which acknowledged the natural law, though without working out its full implications in nature and the cultivation of the soil, possessed a system of distributed ownership, however unequal, in the home, in the workshop, and on the land. The family farm and the small workshop supplementing it, that is to say, are more likely to practice husbandry under the natural law than the manager of a larger unit who delegates authority and loosens responsibility . . .

Husbandry, of course, is not confined to cultivating the soil, whether as farmer or gardener. But as cultivating the soil is the first and most important of all civilized activities because the life of society depends on it, and agriculture is the feet of the commonwealth, we naturally and properly think of husbandry in that connection. But the fullest use of our powers in other avocations, so long as they *are* vocational, is also a kind of husbandry. By so being, it follows that its total meaning is incomplete without allowing for its ethical and aesthetic associations, apart from the religious ones, implicit in its adherence to the natural law. It is needless to labour the point that the exploitation of the soil for cheapness or profit alone is ethically unsound. The 'life' of the soil, which is its organic humus, has logically retaliated upon the inorganic methods (namely, excess use of machines and chemicals) of this acquisitive and financial farming by dying, that is to say, by the dust bowl. To flout the natural law is in the end suicidal. Thus the laws of animate nature demand that man's desire to utilize them shall be governed by an approach which is implicitly ethical. To watch a craftsman persuading and humouring a piece of wood into a certain shape by taking into account its properties reveals both his harmony with nature and his unconscious acknowledgment that man's relations with nature have, in order to achieve practical results, to be moral. All true husbandry expresses a kind of reverence in its manipulations of natural substances, while the 'right' and the 'wrong' ways of doing things, recognized by the older type of labourer in his jobs about the farm, have an indirectly moral bearing.

The ethical and aesthetic aspects of husbandry unite in the term, 'a way of life.' For a way of life means both a pattern of behaviour and a certain rhythm of being in man's intercourse with the earth. We observe, for instance, that when the natural law was operative in agriculture, the country speech, the songs and rituals, the objects made, and the buildings all obeyed another law, the law of beauty. Yet beauty was the very last thing that the pre-industrial countryman tried to capture. The buildings, the songs, the artefacts were made for quite other purposes than for beauty's sake, much less for ornament. Yet beauty is a property of them all, and it appeared as a kind of bonus upon the good work well done for the real needs of the community. It was an unfailing certificate for good husbandry. This something in the universal scheme of things which thus signals its approval of the good works of man has been withdrawn from industrialism and from the predatory power-farming that has hitherto accompanied it.

The Natural Order (1945:6-10)

The Use of Land

QUALITY in diet is very closely related to land-use. It is plain, for instance, that local consumption means bringing the people to their food, and not, as now, bringing their food to the people; and this implies a re-settlement of the land on a regional basis. "Nutritionists", again, are agreed as to the incomparable superiority of fresh over prepared foods, a condition best fulfilled by the natural and dairy foods, fruit, vegetables, eggs, butter and cheese. Where else but from the resources of our own land are we in the future to obtain butter and cheese? The increased turnover would profoundly modify and fertilize existing land-use by, as we have already said, the return of the by-products both for stock-feeding and enriching the land.

But ecological interdependence does not stop there. The return of butter- and cheese-making to the farms would give just the right opportunity for multiplying our pig-population from 2 to 16 million. . . . The pig is one of the most economical of animals for converting wastes both in land and food into protein, while skim milk and whey are ideal foods for young porkers and baconers. Chat (small) potatoes, rye and maize, fodder-beet and barley are other unexacting pig-foods, while a good sow should produce twenty pigs a year. Hence, as we shall continue to stress, pigs are a vital link in the chain between grass, cow, milk and cheese or butter, not to mention their rich manurial value for the fertility of the land.

But quality in food has a direct influence not only upon the quality of land but upon the quantity in land-surface. If, for instance, the staple of life were not robbed of the wheat-germ from the national loaf to increase the profits of the millers, but left in the grain of wheat where nature put it, the whole-grain loaf would enable us to increase the food-value of our wheat acreage by 20%, at the same time trebling or quadrupling the nutritional value of every loaf. More, since whole-grain flour, whether white (by lessening the bran content) or brown, can only be ground by the stone-grinding and not by the industrialized roller mill, the re-conditioning of our windmills and watermills, of which a few only survive as picturesque ruins is highly desirable. Such mills, if only as auxiliaries, and with the aid of turbines, would automatically regulate the current of our rivers, sweeten the waters and check the growth of weeds and rushes. For we can no more afford to waste power, than food. The disingenuous excuse made by the millers that the extraction of the wheat-germ means more feeding-stuffs for beasts is easily put aside for the evasion it is, when we consider how enormously we could increase lucerne as a crop in this country and ensilage for winter fodder, not to mention sugar-beet tops.

The more we intensify, by the avoidance of monoculture and the extension of mixed farming, the productive powers of the farm acreage already in cultivation, the less onerous becomes the task of reclaiming wasteland and restoring built-over land to its proper purpose. Heaven knows there will be enough to do for those missing, forsaken, urbanized, deadened and violated acres, even were we to elevate the whole of our cultivated acreage to the high standard of the English garden. But this standard would at least leave us with a margin; we should not be attempting the impossible. Some of the built-over land cannot be re-converted without almost superhuman effort. But then, what are machines for, if not to make us superhuman? The question remains – how much of it can we afford to leave alone without being half-hearted in approaching our goal of self-sufficiency? And how much more open land can we afford to hand over for non-agricultural use in the future? The answer should work itself out as we go ahead in our creative endeavour, and a new Domesday Book, as suggested by Sir George Stapledon, would be of invaluable service to that end . . .

As for the sprawl of factories over our once-smiling countryside, many of these will be put out of use as industry contracts instead of expanding, and its demands on space become less. For a twentieth-century industry, however prosperous, need not sprawl, and if an individual factory should need more room, then let it grow upwards, not outwards. We must let the procreative blessings of air, light and fertility into the smothered land on which the factory-slums stand. Forestry, the marginal lands and the water-supply we have already reviewed, however inadequately from lack of space. As to the last, we need hardly add that the trapping of rain-water by an inter-linked system of storage tanks should be universal, while the extravagant town-consumption of water may even then have to be curtailed. The re-settlement of the land by *men,* is the most vital of all the great works that lie ahead.

As for parks and recreation grounds, are they any the less recreative, less desirable, less satisfying if they grow good crops and feed good beasts as well? Or if we add herb-gardens, vine-yards and orchards to them? The old craftsmen whom our civilization liquidated could have reminded us of the gracious interplay between use and beauty. Orchards, of course, border many of the French roads, but in this country, they say,the fruit would be stolen. Our answer to this is that a nation of thieves does not deserve to become self-supporting or indeed to survive at all. Passing on to the monster concrete stadia for dog- and motor-racing, we are reminded of the old Spanish proverb – "Take what you will, said God; take it, but pay for it!" They are modern versions of the Roman amphitheatre, pandering to the passions of a dispossessed proletariat which, like ours, was once rooted to the soil. Horse-racing studs and stables could certainly be reduced without hardship too heavy to be borne, while golf-links could be confined to the most sterile types of land and the majority of roads linked to the land by a more regional distribution of them. As for the Fat Boy cities, they will slim themselves as industrialism inevitably yields precedence to agriculture, and becomes really efficient; and as the countryside fills up. Whatever the pangs of transition, it is hardly a tragedy of the first order that urbanism should be transformed into urbanity.

All these measures, or most of them, will be carried out if the finger of necessity touches the heart and soul of Great Britain. What has been set out in so prosaic a fashion will on the contrary appear to many as a Utopian vision, and we are rightly tired of Utopian dreamers. But that the present industrial structure can be preserved intact, and that the ingenuity of the technocrat is able by his charms to call up a magic ship for carrying us over a sea of troubles – this is a fantasy wilder than any Wonderland.

In the history of every nation there comes a time when it has to choose its future destiny or perish of its own stupidity and inertia. It should be plain by now that, having erred from the path of wisdom, we are in danger of being lost for ever. But the alternative still remains – namely, "Tomorrow to fresh woods, and pastures new".

Prophecy of Famine (1953:111-15)

The Organic Way of Life

IT is not to be denied that modern society has made giant strides of discovery. But it is equally clear that it has mainly failed to apply the new knowledge for the benefit of mankind. The failure is so pronounced that it is dragging Western civilisation nearer and nearer to some fall like Lucifer's. Perhaps it can be summed up as the loss of a design for living. To be destroying the earth – birds, beasts, fishes, vegetation and, most of all, soil – in order to make money out of it, and for nearly every country in the world to be fighting either within itself or with other countries, does not make sense. To be producing goods to perish at once in war and in mere quantity with no customers or no money to buy them does not make sense. Common sense is the corner-stone of a design for living. But it is only the cornerstone, nor can the design be restored by abandoning the new knowledge and returning to the past. We cannot put the leaves back on the tree in December. What we can do is to find once more the living truth that crowned it with leaves and fruit.

Of that living truth, the organic way of growth out of the fundamental realities is the beginning. On the material side, a self-supporting and organic agriculture is essential because there is now no other way of feeding the peoples of the world with food enough. By food I mean nourishment from which the vital properties have not been removed nor sacrificed by modern ingenuity in evading organic laws for profit's sake. There is no other way of maintaining the fertility of the soil, that is to say, of passing on human life from age to age. There is no other way of building up resistance to deficiency diseases in man, beast, plant and soil. To support ourselves by organic methods would lead us out of the economic wilderness by opening up vast new fields of employment in the home-market, by making use of local materials now wasted and by diverting trade and finance to national needs with which neither of them has for many years had anything to do. An organic self-help cannot exist without the "just price."

But the spiritual side is just as real and urgent, and the "just price," our practical need, was once a religious precept. Mr. Montague Fordham has well said that "the agricultural problem" is not primarily agricultural nor even political. "It is a problem of statesmanship applied to national life." The organic way is also a means of developing character, independence, ownership, whether co-operative or individual, and a right relation with nature. These are implicitly spiritual needs and an indispensable part of a design for living. If fulfilled, they would at once correct the mechanistic bias of the age. The organic way is, in short, the only one that can bring us within the influence of those everlasting principles that govern mortal life. Taken as a whole, modern urbanism is indifferent or hostile to religious principles. But Germany has given the world an exampl of what happens to a nation when it defies and murders them. We palter with them by turning values into valuations.

The Wisdom of the Fields (1945:251-2)

The Organic Whole

ONLY a few days ago, I spent the afternoon with Samuel Rockall in his workshop looking out on the common zoned in beech-woods. Both he and his son were hard at work on the chair-legs; the one at his chopping-block, the other astride the draw-shave horse. Busy and yet in deep repose, occupied in a score of jobs and yet a multiplicity in unity, such is the life of Samuel. In seeking him out I pass not merely from one place to another, even from one world to another, even from discord to order and from confusion to cohesion. In some mysterious way I pass out from the hallucinations of world-sickness into a speck of reality which is so extraordinarily stable as to defeat even time. I do not mean that Summer Heath will always have its Samuel – in a few years it may well be a bungalow village or a wilderness. What I mean is that the man's organic relation to his natural environment seems to me to be final. Given those woods and that heath – Samuel's life in them and work from them are what theology used to call "the will of God" or "the word made flesh". The natural scene as particularised at Summer Heath was waiting for its Samuel, its king and servant. He arrived, and there he is, the predestined, the meaning of the place, its human expression, its being made manifest in the man. I do not know how else to put it.

Samuel produced a shovel. Forty years ago he had bought it from a travelling tinker for fourpence. Every day for year after year he had gone out on the common with it to pick up what casual offerings from pony, cart-horse, sheep or goat he could find. This promptly went into the compost pit of his garden. In time, the shovel became worn to two-thirds of its original length and Samuel found a bit of metal off a telegraph post blown down by a gale. He hammered and riveted it to the shovel, and the very next day, while he was shovelling the sawdust of his garden-path for fuel, he found a sixpenny piece under it. This, as the Suffolk yokels say, "wholly" pleased him; now his shovel had not only paid for itself but at compound interest. It seemed to me that the fairies, approving of Samuel, must have slipped that sixpence under the sawdust. "I wouldn't", beamed Samuel, "sell this shovel for half-a-crown." Of course not: it was more worth to him than if it had been made of gold. There, in terms of a plain shovel, is the peasant notion of possessions. He makes everything he has go ten times further than another man would do with things ten times the value. Everything the peasant possesses is also ten times itself because of the wealth of associations gathered round it.

We went into the garden. There was a marrow bed which was non-pareil. Samuel expounded the mystery. The roots of the seedlings had been planted on the lip of the compost pit, while a small channel had been dug at the side to let off the liquid manure from the household slops. This vitalised the rest of the garden while the marrows luxuriated in the pit. A few steps away was a specimen of his grafting, a bulge of mud painted black – all Samuel's apples originated from crab-tree stock. This reminded him of the fourteen pounds of blackberry-and-apple jam he had just made in the cottage, and that in turn sent him off to fetch a tobacco jar of wild cherry wood and perfect workmanship he had cut and turned for a customer whom I had unwittingly brought him by my former chronicles of the deeds of Samuel. We returned to the workshop, he to his axe and his lathe, I to the Windsor chair he had fitted up from an 18th century design.

I have known Samuel for years, and yet he has always something new to tell and show me. Except in the felling season, he never leaves his cottage, but out of frugality come riches, and out of simplicity flowers novelty without end. Nothing is wasted; everything turns into something else. Just as nothing is lost in Nature, so with Samuel. The trees he has axed come from the wood; they make multiform objects of use and beauty, while the shavings feed the hearth of the home and the fertility of the garden. The transformations both of garden and wild maintain his family and preserve his independence. Trade, family, economy, livelihood, work, utility, ornament, all are parts of one organic whole. Independence and interdependence are woven into a fabric of wholeness and integrity bears its double meaning. Nature is the source of his industry and each gains by the enrichment of the other. The great primaries of life – Nature, the home, the family, the craft, the land – share an intimate and mutual relation without losing their separate identities. Each is seen to be necessary to the other in the fulfilment of an integrated life. Use and beauty here have no quarrel any more than the wild with the domestic. The cottage and the workshop front the heath and the woods, and this is a symbolic presentation of an inward truth that rules and fructifies the daily humdrum life of Samuel the Bodger. But a life thus creative from dung-heap to flower is homespun rather than humdrum when grace lies within its modesty and wealth within its limitations. All men desire the good life: in the agony of frustration they invent grandiose and visionary Utopias. But on Summer Heath it is embodied like a work of art.

It will thus be seen how inevitably my studies in craftsmanship led me on to husbandry. Husbandry *is* craftsmanship, the sum of all craftsmanship, and to try and turn it into something else, a business among businesses, is the way to utter disaster. Farming is essentially a handcraft, as all crafts are, and that is why extreme mechanisation and standardisation lead only to the literal desert. A craft is always flexible, varied, individual, like life, and the self-sufficiency of mixed farming with all its subtle and delicate inter-relations gives the utmost scope to craftsmanship. The machine, as a farmer once said to me, "takes the song out of the job".

The attempt has been made and is being made to turn farm into factory with the added disqualification to the farmer that his craft is being commercialised without the financial bolstering (the 25 per cent *ad valorem* duty, for instance) accorded to other businesses. Without, too, the expert staff of the business firm, since the better accountant and industrialist the farmer becomes, the worse is his husbandry. The more he tries to adapt himself to an artificial economics, the worse is his violation of biological law. The greater his speed, the more broken is his rhythm. The closer he specialises, the more dislocated are his balances between arable, livestock and grass-ley and the poorer his crops. The more he confuses production with fertility, the sooner he loses the latter. The more he absorbs the urban mentality, the less countryman he. The more up-to-date he becomes, the more down-at-heels are his fields. Land is potential life, not raw material to be manufactured, and the poet's vision of it is far more real and practical than the business man's attitude to it as industrial plant. But the husbandman's treatment of it is the most realistic of all because it is based on an intuitive grasp of biological principles, tested by the experience of many generations.

The judgement of the earth is plain. It will have craftsmanship or nothing and to the predatory man it will refuse its fruits. Its udder will turn sour and then flaccid and sterile. The modern economic system is rejected of the earth because it is false. The earth's answer to it is unequivocal: it is soil-sickness, beast-sickness, man-sickness. A banker's earth is sick at heart. And is there no causation in this triple sickness of soil, beast, man? Vast moneys spent on scientific research and public health have not removed this sickness; they have not even discovered its cause, they have not gathered the three sicknesses into one sickness, which is the sickness of the earth.

But health-wholeness-holiness, only the very rarest man of science is aware of this trinity, a three-in-one. Average science will not stop men from preying on the soil as the plumage-traders preyed on birds in the breeding season for the milliners. An acquisitive society is responsible for the sickness of earth, beasts, plants and men. Nature and the spirit are at one in repudiating an acquisitive society, and such is the answer to Darwinian and company promoter alike. But the relation of the Hunza people (whom I referred to as the healthiest in the world) to nature is not acquisitive but, in Lord Northbourne's admirable term, "symbiotic", just as the craftsman's relation to nature is symbiotic. So the wheel comes full circle and it is possible to claim with some confidence that the human approach to the earth the earth most favours is the craftsman's.

Our modern food is bad because it is denatured out of its wholeness – it is, as we say without knowing what we mean, unwholesome. The profit-making motive is uncraftsmanly because it is unbalanced, hoisting the part at the expense of the whole. The primary question the craftsman-producer wants to know about an article is: "Is it good in itself?" not "Does it pay?" or "Will it sell?" Our subsidies on certain crops are unwholesome because the parts of a farm interlock to make the whole greater than the sum of its parts. Our artificials only restore to the soil part of what is taken out of it, and the non-living part, whereas the "rule of return" restores all the waste as the potential of a new and vigorous life. There is a scientific term, "fractionation", which is beyond me, but, if it means "broken into fractions or fragments" admirably describes the scientific mind. A self-supporting husbandry as opposed to the parasitic reliance upon cheap, stale and imported foods produced by monoculture and exploitation is the true aim and the most practical because it represents the concept of an integer. Cut the physical element out of the cultivation of the soil by substituting machinery for human sweat and experience is halved and stunted. Cut the spiritual element out of it, and the heart of the whole is gone. The farmer is speaking in more than quantitative terms when he says that his land is "in good heart".

Remembrance (1942:134-7)

The Still Small Voice

A very interesting controversy has been taking place in a weekly journal, the circulation of which is so small that it is doubtful whether the vital and all-important issue ventilated has reached the countryman whom it directly concerns. The battle has been between a defender of orthodox modernism in agriculture and his opponents, who have maintained that in actual fact he is out of date (the worst thing that in his own eyes a progressive "agriculturalist" can be!), and nearly all the agricultural scientists with him. The contention of these opponents is that we are passing through not only an economic but a philosophic revolution, which must in the near future profoundly modify our entire attitude not only to agriculture but to nature and to life. The view that at present holds the field, but is being severely challenged, is of man as a self-contained entity in himself, whose destiny is to emancipate himself from the bonds of the past. Riding triumphantly upon the inevitable forces of progress, he will by the development of science and technology achieve the conquest of nature and produce ever-increasing quantities of what modern man requires for his comfort, social opportunities and sustenance. In the past man was enslaved to manual labour, often (and this is quite true) under distressing conditions; now he is being freed by the rapid progress of scientific research, one of its more notable successes being mechanization of agriculture.

Progress, in fact, has become largely a technical problem, the solution to all its perplexities being in the brains of specialists and experts. The answer to this flattering concept is, of course, that it leaves out far more than it puts in, that it disintegrates both nature and man by failing to see them as a whole, and that, being purely quantitative, it turns a blind eye to such factors as quality, fertility, health, and all the creative relationships between man and nature. Moreover, it ignores the facts, and the fact is that this narrow and over-simplified concept, so far from having manufactured a better world, has made the present one full of tension, instability, fear, anxiety, neurosis, disease in nature, loss of vitality in man, decline in soil-fertility and the menace of a world-wide scarcity in food. In plain terms, it has broken down and needs a more satisfying concept to replace it.

Nature, as cannot be over-emphasized, abhors a vacuum and, therefore, under all the clamour and pomp and pride of industrialism a new idea is circulating as a still small voice, the idea that man is not the superior conqueror of nature by his own nature but the essential part of a natural order that comprises a great deal more than himself. Man must learn to conform to the laws of that order and to live in harmony with his natural environment, and the cause of most of the trouble of the world is that he has failed to do so. The value of the past is not an antiquarian one at all, but simply that in the pre-industrial ages man, with all his errors, did practise a way of life in an organic relationship with nature. Indeed, one of the most striking phenomena of our scientific age is its almost total ignorance of the biological and organic life of nature, its lack of culture "in widest commonalty spread", as Wordsworth said (and a rooted culture is not necessarily literate at all), and its contempt for all those intangible truths that cannot be contained within the formulae of the laboratory. The past, therefore, is of the utmost value as a link between the pre-industrial ages and the post-industrial age to come. It will have to come, because by defying the laws of nature the present age is proving to be self-destructive.

We should look upon a farm or garden not as forms of cultivation which must conform to industrial canons, and as presenting a set of problems to be solved by technological short cuts, inordinately expensive both to fertility and in the astronomical bills for fertilizers, insecticides, vaccines, disinfectants, machine repairs and fuel and what not; but as a complex society of living creatures both in and above the soil, capable, if they are in good health, of doing for themselves what we spend so much on artificially trying to do for them. We lavish, for instance, a mountain of money on applying synthetic nitrogen compounds, when by encouragement many species of bacteria could capture atmospheric nitrogen for nothing. And this is only one example of what might be done by making peace with nature, by laying aside the aggressive spirit and by seeing wholes to be studied rather than parts to be dominated.

The Faith of a Fieldsman (1951:258-60)

BIBLIOGRAPHY

Saint Francis of Assisi (translation and adaptation of a play of Josephin Peladan), Scribner, 1913.
Letters to X, Constable, 1919.
People and Things, Allen & Unwin, 1919.
A Treasury of Seventeenth-Century English Verse: 1616-1660 (edited by H.J.M.), Macmillan, 1919.
Dogs, Birds and Others, T. Fisher Unwin, 1921.
Some Birds of the Countryside, T. Fisher Unwin, 1921.
Poems about Birds: An Anthology (chosen and edited by H.J.M.), T. Fisher Unwin, 1921.
Untrodden Ways, T. Fisher Unwin, 1923.
Country Essays, T. Fisher Unwin, 1924.
In Praise of England, Methuen, 1924.
Sanctuaries for Birds and how to make them, George Bell, 1924.
H.W.M. (a selection from the writings of Henry William Massingham with preface and notes by H.J.M.), Harcourt, 1925.
Downland Man, Jonathan Cape, 1926.
Fee, Fi, Fo, Fum or The Giants of England, Kegan Paul, 1926.
Pre-Roman Britain, Ernest Benn, 1927.
The Golden Age: the Story of Human Nature, Gerald Howe, 1927.
The Heritage of Man, Jonathan Cape, 1929.
The Friend of Shelley: A Memoir of Edward John Trelawny, Cobden-Sanderson, 1930.
Birds of the Seashore, T. Werner Laurie, 1931.
The Great Victorians (ed. jointly with Hugh Massingham), Nicholson & Watson, 1932.
Wold without End, Cobden-Sanderson, 1932.
London Scene, Cobden-Sanderson, 1933.
Country, Cobden-Sanderson, 1934.
English Country: Fifteen Essays by Various Authors, (edited and introduction by H.J.M.), Wishart, 1934.
Through the Wilderness, Cobden-Sanderson, 1935.
English Downland, Batsford, 1936.
Genius of England, Chapman & Hall, 1937.
Cotswold Country: A Survey of Limestone England from the Dorset Coast to Lincolnshire, Batsford, 1937.
Shepherd's Country: A Record of the Crafts and People of the Hills, Chapman & Hall, 1938.

The Writings of Gilbert White of Selborne (Introduction and edited by H.J.M.), Nonesuch Press, 1938.
English Countryside: A Survey of its Chief Features (introduction and chapter by H.J.M.), Batsford, 1939.
Country Relics: An Account of Tools etc. once belonging to English Craftsmen, Cambridge University Press, 1939.
A Countryman's Journal, Chapman & Hall, 1939.
The Sweet of the Year: March to June, Chapman & Hall, 1939.
Chiltern Country, Batsford, 1940.
The Fall of the Year: July to December, Chapman & Hall, 1941.
England and the Farmer: A Symposium (ed. H.J.M.), Batsford, 1941.
Home, Dent & Sons, 1942.
Remembrance, Batsford, 1942.
The English Countryman: A Study of the English Tradition, Batsford, 1942.
Field Fellowship, Chapman & Hall, 1942.
The Tree of Life, Chapman & Hall, 1943.
Men of Earth, Chapman & Hall, 1943.
Chapter by H.J.M. in *Return to Husbandry,* Dent & Sons, 1943.
This Plot of Earth: A Gardener's Chronicle, Collins, 1944.
The Wisdom of the Fields, Collins, 1945.
The Natural Order: Essays in the Return to Husbandry (edited and chapter by H.J.M.), Dent & Sons, 1945.
Where Man Belongs, Collins, 1946.
The Small Farmer: A Survey by Various Hands (edited, preface, and chapter by H.J.M.), Collins, 1947.
An Englishman's Year, Collins, 1948.
The Curious Traveller, Collins, 1950.
The Faith of a Fieldsman, Museum Press, 1951.
The Southern Marches, Robert Hale, 1951.
Prophecy of Famine (with Edward Hyams), Thames & Hudson, 1953.

H.J. Massingham also contributed to the following books:

Lark Rise to Candleford, Flora Thompson (Oxford, 1945), introduction.
From The Ground Up, Jorian Jenks (Hollis & Carter, 1950), introduction.
The Great Tudors, John Lyly (Nicholson & Watson, 1935).
The Post Victorians, W.H. Hudson (Nicholson & Watson, 1933).

BOOKS ABOUT H.J. MASSINGHAM

There has been no biography or single book about Massingham. After his death in 1952, Rolf Gardiner devoted the Christmas 1952 edition of *Wessex - Letters from Springhead* to tributes to Massingham by himself, Adrian Bell, Edmund Blunden, Arthur Bryant, J.E. Hosking, Lord Portsmouth, and C. Henry Warren.

The Rural Tradition by W.J. Keith (Harvester Press, 1975), a critical study of non-fiction prose writers of the English countryside, contains a chapter on Massingham.